Simiae	Ceboidea (New World monkeys, platyrrhine monkeys)	Cebidae	Cebinae	*Pithecia*	Saki
				Chiropotes	Saki
				Cacajao	Uakari
				Alouatta	Howler monkey
				Saimiri	Squirrel monkey
				Cebus	Capuchin
				Ateles	Spider monkey
				Lagothrix	Woolly monkey
	Cercopithecoidea (Old World monkeys, catarrhine monkeys)	Cercopithecidae	Cercopithecinae	*Macaca*	Macaque
				Cynopithecus	Black ape
				Papio	Baboon, drill, mandrill
				Theropithecus	Gelada
				Cercocebus	Mangabey
				Cercopithecus	Guenon
				Erythrocebus	Patas monkey (hussar monkey, red monkey)
			Colobinae	*Presbytis*	Langur, leaf-monkey
				Pygathrix	Douc
				Rhinopithecus	Snub-nosed monkey
				Simias	Pig-tailed langur (Mentawi Islands langur)
				Nasalis	Proboscis monkey
				Colobus	Guereza
	Hominoidea (apes and man)	Hylobatidae (lesser apes)		*Hylobates*	Gibbon
				Symphalangus	Siamang
		Pongidae (great apes)	Ponginae	*Pongo*	Orangutan
				Pan	Chimpanzee
				Gorilla	Gorilla
		Hominidae		*Homo*	Man

NOTE: Names in parentheses are synonyms for the names they immediately follow. Names separated by commas, but not in parenth are not synonyms.

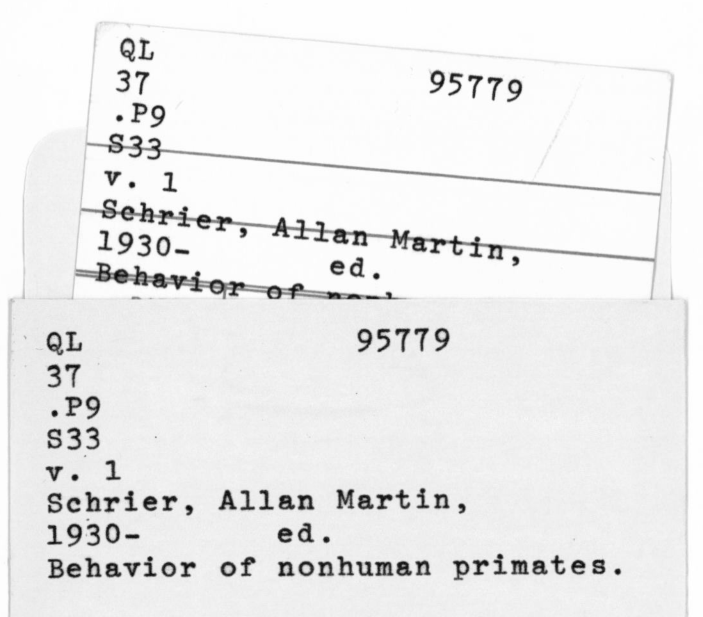

Behavior of Nonhuman Primates

MODERN RESEARCH TRENDS

Volume I

Contributors To This Volume

HAROLD J. FLETCHER
GILBERT M. FRENCH
ROGER T. KELLEHER
MARVIN LEVINE
DONALD R. MEYER
PATRICIA M. MEYER
RAYMOND C. MILES
F. ROBERT TREICHLER
J. M. WARREN

Behavior of Nonhuman Primates

MODERN RESEARCH TRENDS

EDITED BY

ALLAN M. SCHRIER
PRIMATE BEHAVIOR LABORATORY
WALTER S. HUNTER LABORATORY OF PSYCHOLOGY
BROWN UNIVERSITY
PROVIDENCE, RHODE ISLAND

HARRY F. HARLOW
PRIMATE LABORATORY
DEPARTMENT OF PSYCHOLOGY
UNIVERSITY OF WISCONSIN
MADISON, WISCONSIN

FRED STOLLNITZ
PRIMATE BEHAVIOR LABORATORY
WALTER S. HUNTER LABORATORY OF PSYCHOLOGY
BROWN UNIVERSITY
PROVIDENCE, RHODE ISLAND

Volume I

1965

ACADEMIC PRESS New York and London

ACADEMIC PRESS, INC.
111 Fifth Avenue, New York, New York 10003

United Kingdom Edition published by
ACADEMIC PRESS, INC. (LONDON) LTD.
24/28 Oval Road, London NW1

LIBRARY OF CONGRESS CATALOG CARD NUMBER: 65-18435

Second Printing, 1973

PRINTED IN THE UNITED STATES OF AMERICA

To our wives, Judith, Peggy, and Janet

List of Contributors

Numbers in parentheses indicate the pages on which the authors' contributions begin.

HAROLD J. FLETCHER, Department of Psychology, University of Wisconsin, Madison, Wisconsin (129)

GILBERT M. FRENCH, Department of Psychology, University of California, Berkeley, California (167)

ROGER T. KELLEHER, Department of Pharmacology, Harvard Medical School, Boston, Massachusetts (211)

MARVIN LEVINE, Department of Psychology, Indiana University, Bloomington, Indiana (97)

DONALD R. MEYER, Laboratory of Comparative and Physiological Psychology, The Ohio State University, Columbus, Ohio (1)

PATRICIA M. MEYER, Laboratory of Comparative and Physiological Psychology, The Ohio State University, Columbus, Ohio (1)

RAYMOND C. MILES, Department of Psychology, The Ohio State University, Columbus, Ohio (51)

F. ROBERT TREICHLER, Department of Psychology, Kent State University, Kent, Ohio (1)

J. M. WARREN, Department of Psychology, The Pennsylvania State University, University Park, Pennsylvania (249)

Preface

Research on the behavior of nonhuman primates has mushroomed in the last decade. Not many years earlier, primate behavior was a major area of study in only two laboratories in the United States, and only a few isolated field studies had been made. Today, in addition to the pioneering Yerkes Laboratories of Primate Biology and the University of Wisconsin Primate Laboratory, there are primate behavior laboratories in many university psychology departments and government research institutions, and long-term field studies have become almost common. But the wealth of new data on primate behavior is relatively inaccessible to most behavioral scientists. It cannot be described adequately in books covering the wider field of animal behavior or comparative psychology, and the original research reports are scattered through hundreds of volumes of scientific journals.

To make a substantial part of this new knowledge of primate behavior more readily available, we have tried to obtain thorough and up-to-date descriptions of fifteen important research areas. The resulting volumes do not include every study in the literature on primate behavior; rather, each of the chapters attempts to give a coherent, integrated view of its particular area. Some historical background is provided, but the emphasis is on modern research trends. We hope that these chapters will provide new perspectives not only to researchers using nonhuman primates as subjects, but also to other readers—for example, to those who are mainly interested in human behavior but are also curious about man's closest relatives, and to those who study rat behavior but wonder how well rodent results (and rodent theories) can be extrapolated to primates. Studies of nonhuman primates have already been influential in the conceptual development of the behavioral sciences; this influence may well increase as work on primate behavior becomes better known.

We have assumed that many readers would be familiar with the study of behavior, but not necessarily with concepts and techniques specific to primate research. Each chapter either describes techniques and problems fully enough for nonspecialists or provides cross-references to detailed descriptions in other chapters. In addition, Volume I, which concentrates on studies of learning and problem-solving, starts with a chapter that emphasizes methods used in many of these studies.

Nonhuman primates are referred to by their vernacular names in this book. Identification of some animals presents special problems, which are discussed further in the Appendix, but we may note here that the scientific name is also given in each chapter the first time that a particular vernacular name is mentioned. A classification of living primates is presented on the endpapers of each volume. Although very little is said about the prosimians in this book, we believe that this gap in our knowledge is temporary and that the approaches to behavior that have been so fruitful with apes and monkeys will soon be extended to the prosimians.

In acknowledging our gratitude to others, we recognize immediately our double debt to the authors, who literally made these volumes possible. We thank them first for their contributions, which reflect their high competence as researchers and scholars, and second for their cooperation with editors who thought that their specialties might be of interest to nonspecialists and tried to edit accordingly. We are also grateful to a number of other people: Harold Schlosberg, the late chairman of the Brown University Psychology Department, who was an encouraging and understanding friend; other members of the Department, who provided valuable advice on several questions related to their special areas of interest; Kathryn M. Huntington, who tirelessly typed and retyped manuscripts in various stages of revision; Richard A. Lambe and Karen E. Lambe, who were an excellent proofreading team; Vicky A. Gray, who gave editorial assistance in a variety of ways; and Kenneth Schiltz, who labored through the University of Wisconsin's photographic files to find figures for several chapters. Preparation of this book was supported in part by U. S. Public Health Service Grant MH-07136, from the National Institute of Mental Health.

April, 1965

ALLAN M. SCHRIER
HARRY F. HARLOW
FRED STOLLNITZ

Contents

Chapter 3

HYPOTHESIS BEHAVIOR

By Marvin Levine

Chapter 4

THE DELAYED-RESPONSE PROBLEM

By Harold J. Fletcher

Chapter 5

ASSOCIATIVE PROBLEMS

By Gilbert M. French

Chapter 6

OPERANT CONDITIONING

By Roger T. Kelleher

Chapter 7

PRIMATE LEARNING IN COMPARATIVE PERSPECTIVE

By J. M. Warren

Contents of Volume II

Chapter 1

Discrete-Trial Training Techniques and Stimulus Variables[1]

Donald R. Meyer

Laboratory of Comparative and Physiological Psychology, The Ohio State University, Columbus, Ohio

F. Robert Treichler

Department of Psychology, Kent State University, Kent, Ohio

and

Patricia M. Meyer

Laboratory of Comparative and Physiological Psychology, The Ohio State University, Columbus, Ohio

I. INTRODUCTION

This is an account of factors that are known to govern the efficiency with which the nonhuman primates learn in situations that involve a trial-by-trial approach. The variables of interest are the nature of the cues, their modes of presentation to the subject, and the spatiotemporal relations of cues to responses and rewards. There now is a most substantial body of research related to this group of variables, and it has been shown that very powerful effects can be produced by their manipulation.

[1] This review and the research reported to have come from The Ohio State University have been supported by a research grant (M-2035) from the National Institute of Mental Health, U. S. Public Health Service.

We shall deal exclusively with methods that involve the use of discriminanda, and mainly with experiments in which the visual cues are of a fairly complex nature. We propose, initially, to give a brief account of how our present methods came to be, and we then shall follow this with a discussion of the properties of current apparatus. Then we shall consider what now appear to be the major problems that confront investigators who propose to build and operate sophisticated training instruments.

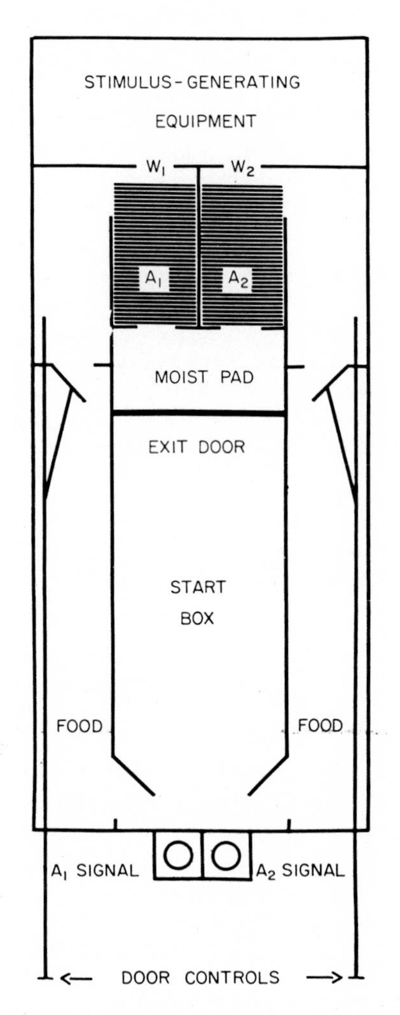

Fig. 1. Floor-plan of the Yerkes-Watson apparatus used by H. M. Johnson. (After the floor-plan of Johnson, 1914, but with changes indicated in the text of Johnson, 1916.)

II. SOME CLASSIC PROCEDURES AND RESULTS

The earliest experiment of lasting worth to be performed with non-human primates was the work of H. M. Johnson (1914, 1916) on the visual acuity of cebus monkeys (*Cebus*). The study was conducted with the Yerkes-Watson box, a leading apparatus of the day, and one that was designed to be adaptable for use with any kind of animal subject. The discriminanda were displayed on a wall near the entrances of alleys leading to goal compartments that were placed on either side of the start box (Fig. 1). Though this apparatus was convenient to use, investigators had some problems with it; animals did not learn quickly, and one reason was delay of reinforcement after choice. Johnson, nonetheless, was able to establish that the limits of acuity for cebus monkeys were about the same as those for men when comparable conditions were employed.

Subsequent investigators rarely chose to work with locomotor training situations. Rather, their contrivances were more specifically adapted to their primate subjects. In the very early years, as we might expect, primate training methods varied greatly; some approaches were informal and involved the use of relatively simple apparatus. One extreme is

Fig. 2. Madame Kohts presenting a color sample to her chimpanzee. (After Kohts, 1923.)

illustrated by the work of Kohts (1923), who studied matching in a chimpanzee (*Pan*): the animal was asked to pick an object from a tray after it had seen or felt a sample (Fig. 2). Many of the questions that have been of some concern to apparatus builders ever since were tentatively answered, and quite accurately, by this pioneering primatologist.

Generally, the most efficient methods were the ones in which the least control was exercised. Neither apes nor monkeys did particularly well when they were trained in formal situations, and a first example is provided by the work of J. A. Bierens de Haan (1927) with two pig-tailed macaques (*Macaca nemestrina*) and a mangabey (*Cercocebus torquatus*). This investigator used a glass-fronted case mounted on a wooden choice compartment (Fig. 3). The discriminanda were displayed within the case, and were solid objects or their pictures. Food rewards were placed within the underlying box, behind a pair of covered entry ports, and the monkeys made a choice by reaching into one of the ports. This arrangement was so inefficient that a pair of cones whose heights were 16 and 11 cm were not discriminated by the pig-tailed macaques within 1,250 trials.

Gellermann (1933a, 1933b) had much the same experience when he embarked on a study of performances of chimpanzees and children in another kind of visual-learning situation. For this work, he used a pair of food boxes mounted on a long, narrow shelf. Patterns were placed in a frame between the boxes in the first of his arrangements (Fig. 4). Under these conditions, two young chimpanzees completely failed to

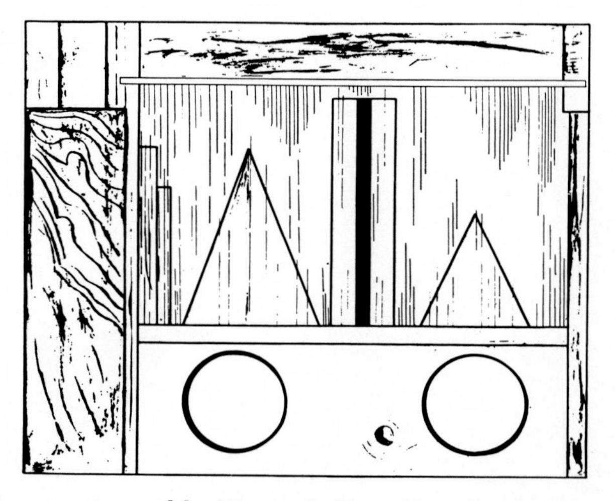

Fig. 3. The apparatus used by Bierens de Haan. Note the placement of the cones behind a sheet of glass and over the response apertures. (After Bierens de Haan, 1927.)

learn a pattern discrimination, but they learned when the patterns next were placed adjacent to the lids of the food boxes. Even then, their rates of learning were extremely slow, and it still was possible to argue that extended training would be required if a well-controlled method eliminated extraneous cues.

FIG. 4. The apparatus used by Gellermann. (After Gellermann, 1933a.)

Klüver (1933), in the meantime, had developed an approach that offered one solution to the problem. His classic method of equivalent stimuli let him make the fullest use of an investment in initial training of an animal. In his applications of the method to the study of perception in the lower primates, this great innovator used a number of techniques, one of which is shown in Fig. 5. In the apparatus shown, the monkey was

FIG. 5. One of the varieties of Klüver's pulling-in apparatus. (After Klüver, 1933.)

required to pull in one of two food boxes, basing its choice on cues provided by a pair of cards attached to the boxes. The method of equivalent stimuli involved initial training with a pair of cards and tests of the reactions of the animals to cards that differed from the training stimuli.

Klüver's method has been used by many workers since, but few have worked with pulling-in techniques. Although they seemed efficient with respect to trials, these techniques were cumbersome to use, and the methods that replaced them share the attribute of letting the experimenter sit.

The first such apparatus to combine efficiency, convenience, and control of variables was one that Spence (1937) developed in connection with his work on theory of discrimination learning. Food boxes similar to those that Klüver used were featured in the Spence apparatus, but they were placed upside down and attached by hinges to a tray that moved freely on a set of wheels (Fig. 6). The discriminanda were sheet-metal forms that could be attached to the boxes. The study was conducted with chimpanzees that responded through the mesh of a cage. All the

FIG. 6. The type of apparatus used by Spence (1937), but here equipped with five unmarked boxes. The experimenter holds a stick used in moving the tray. (After Yerkes, 1943, plate 39.)

operations of preparing for a trial were hidden from the subject by a screen, which was raised when the tray was ready. Then the tray was moved to where the ape could reach the two discriminanda with its fingers, and a push would tip a food box backward to expose whatever food it might have concealed.

Spence observed that chimpanzees could learn to push the boxes in no more than half a dozen trials, and that they could learn discriminations of two forms within about a hundred presentations. He gave several reasons for this favorable result and stressed the fact that he displayed the forms in such a manner that these cues were distinctly set apart from those that had no relevance for learning. He believed that it was important that his apes were required to push directly against the forms, and that these responses were contiguous in time and space with presentations of rewards. Much of what has since been learned about the properties of training apparatus for the primates can be viewed as an extension and a confirmation of this set of simple principles.

Spence's combination of a cage, a moving tray, and a screen is found in almost every widely used, successful instrument that has since been devised for work with primates. Boxes, on the other hand, have largely been replaced by foodwells that are cut into the tray, and the forms and patterns that were once attached to them simply serve as covers for the foodwells. Such a tray is termed a Klüver (1937) formboard, and it is simplicity itself, but it is, without a doubt, the most important tool that primatologists have yet devised. With it, Harlow (1942) undertook the program that has led, above all others, to our modern concept of what monkeys can be taught to do if they are trained with methods that are suited to the monkey.

Harlow's justly famous, widely "modified" Wisconsin General Test Apparatus (WGTA; Harlow & Bromer, 1938; Harlow, 1949) represents the happiest of marriages between the Klüver formboard and the Spence arrangement (Fig. 7). The experimenter faces the caged subject across a table, which supports the formboard (often called the test tray or stimulus tray) and a superstructure holding screens. Typically there are two screens, an opaque one that can be lowered between the subject and the tray to prevent the subject from observing preparations for a trial, and a one-way-vision screen that can be lowered between the tray and the experimenter to prevent the subject from observing the experimenter during a trial. This simple apparatus has been the workhorse of behavioral research with primates for a quarter of a century, and the present indications are that it will have a place in such research for years to come. The WGTA, as we shall see, has certain rather striking limitations, but in certain circumstances it remains the most efficient apparatus that we have.

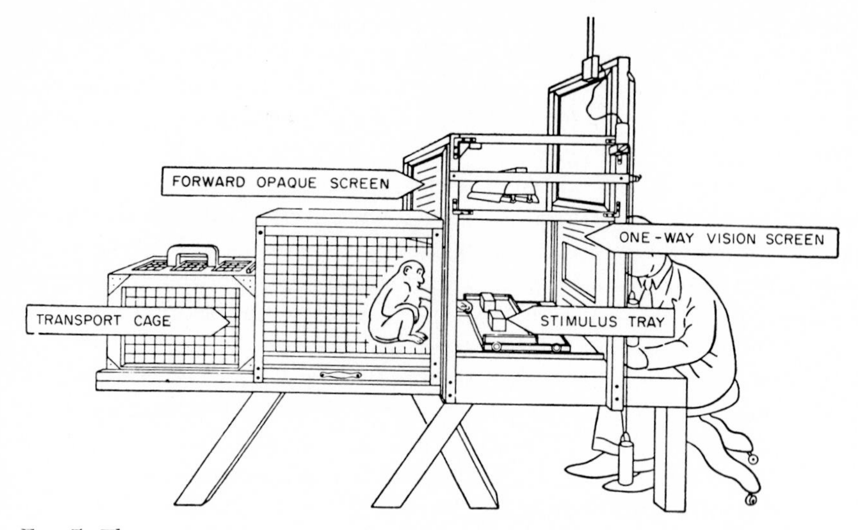

FIG. 7. The circa-1949 model of the Wisconsin General Test Apparatus. (After Harlow, 1951.)

III. ON WORKING WITH THE WISCONSIN GENERAL TEST APPARATUS

There are now so many different WGTAs that we cannot consider all their features, but there are some fairly common things about their use that we believe deserve description here. Some of these are founded in empirical results, but many are the product of beliefs that those who have employed the method gradually have built into a WGTA mystique. Rarely are these points of interest in the general sense, but they can have certain consequences for the smoothness of the planning and the execution of a test of some important issue.

The first and most important steps in all of this research are those by which the animals are tamed. It is simply foolish to attempt to train a monkey on a fairly complicated problem without first adapting it to people, apparatus, and the routine of the laboratory. In our view, the most effective way of doing this, although it is a tedious procedure, is to start by placing several monkeys in a cage that also will accommodate a man. A waterproof raincoat and a hat are *de rigeur,* and the tamer should have steady nerves, but we know of no one who has ever been attacked in this situation by young, wild rhesus monkeys (*Macaca mulatta*). Rather, their initial contacts with the man are reasonably likely to be stepping on his shoulder or his head as they progress about the upper reaches. The use of groups facilitates the process at the start (Weinstein, 1945).

Individual taming can begin as soon as any subject will, without hesitation, eat food from the floor of the cage. An effective method, at this point in the procedure, is to offer food from the hand. After several days of this, a progressing subject can be trained to open one's fist. Such a monkey then is ready to be trained to enter "transport" cages placed upon the floor. These are small cages with sliding doors at each end, which are used to transfer monkeys from their home cages to the test apparatus. This step of taming is facilitated if both doors are open during the initial stages. One can begin by putting food within the cage or giving it by hand within the cage; when the monkey enters freely, one door is replaced, then the other, and the taming process is complete. One important dividend of training of this kind, apart from ease of handling the subjects, is that monkeys that escape in strange surroundings seek the transport cage as a familiar haven.

The above procedures, while they may take several weeks, yield a group of subjects that can be quite readily adapted to the WGTA. Here the best indication of success is cessation of defecation in the apparatus. Almost any change in a procedure will induce substantial dirtying of the equipment, but a subject that has been ideally tamed will not persist in this beyond a day or so.

With a naive subject, some experimenters begin by feeding subjects from the hand or tossing food into the WGTA's restraining cage with the screens raised. The animals then are offered food on the surface of, and within the foodwells of, the test tray. Next they are adapted to manipulations of the opaque screen, and they then are finally shaped up to push aside a single block that covers a foodwell.

Even monkeys in which such investments have been made may not be too consistent in their work. Noises are, without a doubt, the single most important general environmental variable. Thus, though they habituate to a repeated sound (Butler & Harlow, 1956), occasional stray noises have effects. Monkeys that are separated from a colony will listen for its characteristic sounds, and if these are audible, the animal will spend a great deal of its time in answering. Cross-talk of this sort is also commonly observed between two individual monkeys if concurrent testing is performed within a suite of laboratory rooms that have not been sound-treated.

Deprivation schedules, on the other hand, seem to be of little consequence (Meyer, 1951a). Thus, provided that one uses, for rewards, foods that rhesus monkeys markedly prefer, it does not appear to matter greatly when they last were fed a nonpreferred standard ration. Monkeys of this species that have had a dish of biscuits just an hour prior to being tested will perform about as well for good rewards as will monkeys that have not had standard rations for 2 days. Miles (1959) has found that deprivation times in the range from 1 to 20 hours are without effect on discrimination learning in the squirrel monkey (*Saimiri sciureus*) (Fig. 8). Therefore, more attention has been given to the question of what a monkey likes to eat than to the assessment of the level of starvation that will make it work for anything.

Measurements of preferences for different kinds of foods have been obtained through paired comparisons (Harlow & Meyer, 1952), and estimates obtained at one time can be reproduced in tests conducted 3 months later (Fig. 9, left). It has also been established, by the same technique, that scale values vary in a linear relation to logarithmic changes in amount (Fig. 9, right). Further, Fay *et al.* (1953) have

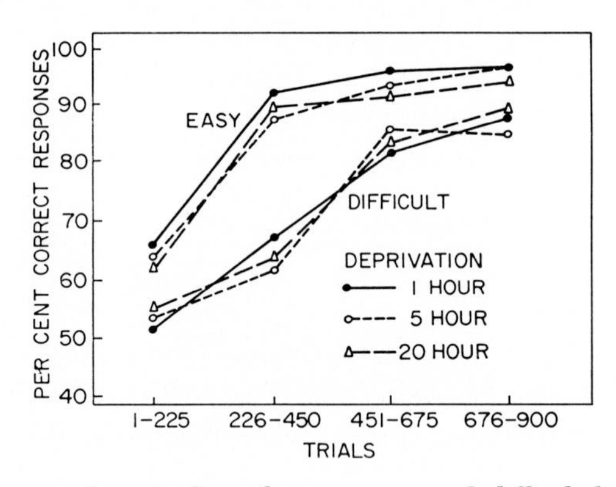

FIG. 8. Performance of squirrel monkeys on easy and difficult brightness discriminations after deprivation periods of three different lengths. (After Miles, 1959.)

shown that the preferences relate to the ratio of amounts of two kinds of food even though the foods also differ in quality as judged by monkeys. That is, probabilities of choice between one-quarter peanut and two pieces of potato are matched almost perfectly by probabilities of choice between one-half peanut and four pieces of potato. Two days' deprivation, incidentally, does not affect the monkeys' choices of these foods; hungry monkeys show no bias toward acceptance of the larger, but the less-preferred, reward. Berkson (1962) has found that food preferences of gibbons (*Hylobates lar*) are similar in most respects to those observed for monkeys and for juvenile chimpanzees (unpublished data of Mason & Saxon, cited by Berkson, 1962).

Other work has shown that, within limits, reward context makes a difference. For example, Meyer and Harlow (1952) found that differences in quantity of a preferred reward had very little effect on delayed-response performances of rhesus monkeys until the animals had had considerable practice. Further, there is evidence (Meyer, 1951b) that such discriminations of reward levels do not generalize. Thus, a group of monkeys that had performed differentially for different quantities of a preferred food in delayed-response testing did not perform differentially on discrimination-reversal problems for two different quantities of food until practice had been given with that reward differential in the new task situation (see Chapter 2 by Miles).

A reasonably safe approach, accordingly, involves the use of reinforcers that, for any given monkey, will sustain that animal's responding. Leary (1958), for example, found that levels of performance by rhesus monkeys on concurrent object-discrimination problems (see Chapter 5

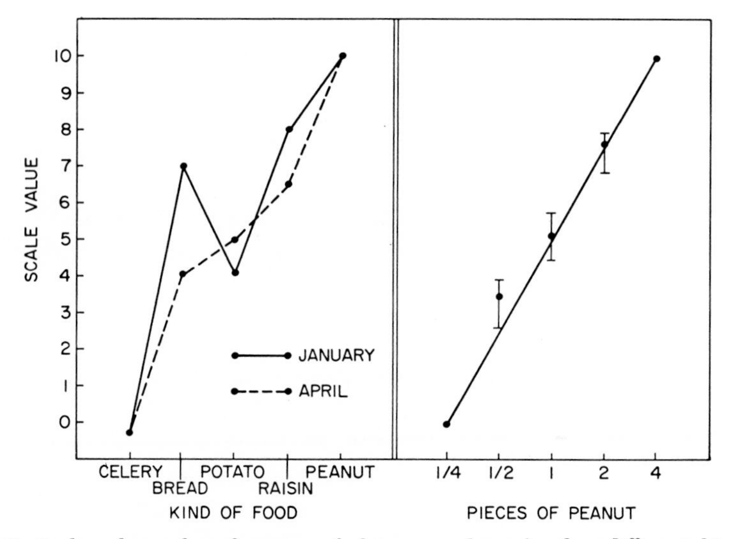

FIG. 9. Scale values of preferences of rhesus monkeys for five different kinds and quantities of food. (After Harlow & Meyer, 1952.)

by French) did not differ for two groups when one of these was given four times the reward of the other. Schrier (1958), who gave groups of rhesus monkeys 1, 2, 4, or 8 food pellets for performance on discrimination tasks, found that the 8-pellet group was somewhat better, but that the others were the same. Differences can be obtained, and were in both these studies, with a smaller difference in reward, but such differences depend upon each animal's experiencing several rewards (see Chapter 2 by Miles).

Less is known about such factors as the work demanded during execution of a choice. Davis's (1956) experiment on reach-work is of interest in that anyone who uses a formboard must decide how far it is to be from the bars of the monkey's restraining cage during a test trial. Davis first determined the maximal reach of each member of a group of rhesus monkeys, and then examined the effects of setting the test tray at different fractions of the total distance. Variations of this kind had very small effects on three different measures of performance. Most investigators who attempt to standardize the variable prefer the longer reaches, but the probability of failure to respond goes up very rapidly for reachwork levels in excess of 90% of maximum reach.

The tray, of course, is commonly withdrawn from reach of the animal when a given trial has been completed, and an opaque screen is lowered so that preparations can be made for the next trial. In the early studies with the WGTA, the operator typically arranged the stimuli and the reward and then advanced the tray before the opaque screen was raised. Later, it became the custom to withhold the tray until the opaque screen was raised, and thus to force the animal to wait a time before it made responses to the group of objects. These manipulations meant that one-way-vision screens, which otherwise would be completely lowered, now were lowered only to a point where they concealed the operator's face and upper body.

Several of the later versions of the apparatus, for example, Riopelle's (1954) and Schrier's (1961), have employed transparent screens to enforce the same brief delay after the opaque screen is raised. The transparent screen is then raised to allow a response. No one knows exactly what the observation times should be for any given class of test, but that observation times can help performances is nonetheless a widely held belief.

These, then, are the more important incidental points concerning training methods for the primates that may not be too familiar to experimenters who have not used WGTAs. They are points that have to be considered in planning a study with this type of apparatus, but to us the striking thing is that they leave a lot of freedom for design and execution. But, though one can often, though not always (Gross, 1963; see Chapter 4 by Fletcher) disregard such variables as deprivation schedules, those that we shall next consider have been shown to be pervasively important. Basically, they have to do with problems that arise when one must choose a group of stimuli; we shall see that things that shouldn't matter often do, and do so to astonishing degrees.

IV. SOME FACTS ABOUT THE PROPERTIES OF OBJECTS

For 20 years, the formboard methodologist has been confronted with what seems a simple problem, but this problem still exists despite the large amount of work that has been aimed at its solution. Basically, the problem has to do with why it is that complex, common objects of the sort that Weinstein (1941, 1945) first employed in his experiments on matching are such splendid cues for rhesus monkeys (Fig. 10). Harlow was the first investigator to attempt to understand the nature of such objects, and we will begin with an account of the results that he obtained in his initial studies.

In the first of several in a series of experiments, Harlow (1944) trained six rhesus monkeys to a criterion of 23 correct in 25 responses on each of five discrimination problems. Afterwards, he found that 15 object-

FIG. 10. Examples of the objects used by Weinstein in his work on matching to sample by rhesus monkeys, and which are quite similar to those in common use in WGTA experiments. (After Harlow, 1951.)

quality-discrimination problems all were solved essentially at once; with a criterion of 10 out of 10, the mean error score was 1.3. The animals were then trained on 10 discrimination-reversal-learning problems, which they solved with a mean of only 2.1 errors per problem. These results, obtained with complex objects, were, to say the least, spectacular; no one prior to this had even found a way of teaching habits quite so rapidly.

Harlow's second study in the series dealt with matters that are not related to our task, but his third (Harlow, 1945a) was given over to comparisons of several classes of discriminanda (Fig. 11). Some of these were wooden forms that were thick or thin, being cut from two-by-fours or plywood. Some thin forms were held upright by small, concealed supports; others merely lay upon the test tray. Others were affixed with glue to white wooden wedges, and such wedges also were employed for the display of black or colored forms painted on the centers of their upper surfaces. The outstanding finding was that cues upon such wedges are much less effective for rhesus monkeys than are these same cues when they are placed directly upon the tray.

FIG. 11. Examples of discriminanda used by Harlow to assess the consequences of the thickness of an object, its orientation, and its background. (From Harlow, 1945a.)

Harlow's fourth experiment (1945b) was hence devoted to a further study of the wedge effect. Here he used a set of 12 discrimination problems for which the cues were cut-out wooden forms that either were placed directly on the tray (object discriminations) or else were glued to wooden wedges (pattern discriminations). The wedges were either uniformly white or white with very prominent black borders, but, within a given pair, the wedges were the same, and hence did not provide an extra cue (Fig. 12). Once again, the learning rates for pairs of cut-out forms were faster than for forms glued to wedges, but the nature of the wedge did not appear to make a difference to the rhesus monkey subjects.

Harlow's fifth experiment (1945c) was undertaken with a group of experimentally naive subjects, nine rhesus macaques and three bonnet macaques (*Macaca radiata*). They were pretrained with five objects presented one at a time for many trials. By pushing aside these single

Fig. 12. Examples of discriminanda used by Harlow to determine whether borders change the effects of mounting pairs of objects on white wedges. (From Harlow, 1945b.)

objects, the monkeys found food in the left and right foodwells equally often before they were given their first discrimination problem. Then they were trained on 12 problems, 6 of which involved objects; the remaining 6 involved patterns (cues on identical bases) (Fig. 13). The mean error scores with objects proved to be about a third of those obtained with patterns. When objects were used, the subjects often solved the problems after single presentations. The extensive pretraining contributed to this efficient performance (Harlow, 1959, p. 525).

Fig. 13. Examples of the free-standing objects and wedge-type pattern discriminanda used by Harlow in his work with naive monkeys. (From Harlow, 1945c.)

Harlow, in his sixth and final study of the series (1945d), worked with differences in form and color. He presented objects in the first part of this work, and patterns in the second (Fig. 14.) Here he found that color problems were much easier than those involving differences in form, and that the addition of a difference in form had no effect if color cues were present. This result was puzzling because the notion was that the efficiency of object-discrimination learning is, in part, a function of the multiplicity of differences between discriminanda. Yet, when Meyer and Harlow (1949) asked the question once again, this time for oddity learning (see Chapter 5 by French), the outcome was essentially the same: color cues alone were just about as good as cues of color, form, and size combined.

There were two main problems that emerged from this research, and lack of additivity was one; the other was the fact that the effectiveness of cues is markedly affected by a base. We should note that Klüver (1933, pp. 172–176, 190) had encountered this effect when he presented cues on two cards; now, however, it was also known that the effect ap-

FIG. 14. Examples of the object and pattern discriminanda used by Harlow to assess summation of form and color cues. Note that each object is mounted on a small base of the same color as the object. (From Harlow, 1945d.)

pears when cues are pairs of solid objects. What was puzzling here was that a free-standing object stands upon a tray that, in itself, surrounds the object with a group of cues that are as neutral as the background stimuli of wedge bases. Nonetheless, attachment of the backgrounds to the cues-to-be-discriminated made a difference, and it is but recently that we have come to know the fundamental reason for the difference.

V. ON BORDER CUES AND ADDITIVITY

A number of the next experiments to be devoted to analyses of cues were carried out with cardboard plaques whose surfaces were used for the display of thin materials. It was hoped that some convenient substitute could be developed for the bulky solid object, for there was no proof that three-dimensionality facilitates discrimination learning. The first of several studies in this general area was carried out with rhesus monkey subjects (Harlow & Warren, 1952). The investigators covered 3-inch cardboard plaques with squares of paper cut from magazines. The resulting "random patterns" thus were highly varied with respect to color and to form, but they were of equal weight, identical in size, and all contained within a common format.

As discriminanda, "random pattern" plaques were neither very good nor very bad; they were much superior to plaques with central cues, but markedly inferior to objects. Therefore, they were never used extensively in work requiring many different stimuli, and the patterns that were next examined with the cards were much more analytic in their nature.

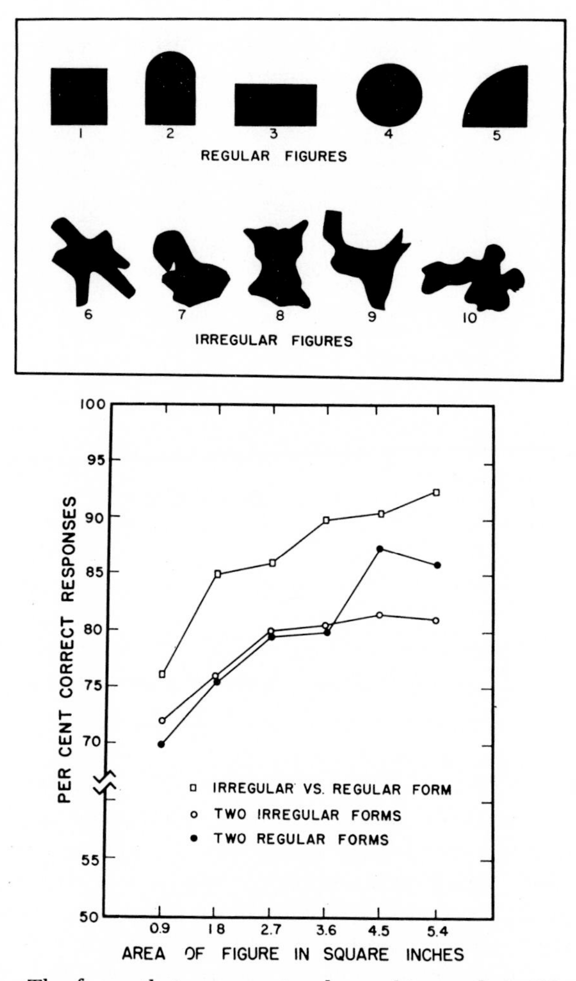

FIG. 15. Top: The forms that were centered on white cards in Warren's study of the effects of size and contour on discrimination learning. Bottom: The principal results of the study. (From Warren, 1953b.)

One could wonder whether forms in "random pattern" plaques contribute to their differentiation, for, from Harlow's other work, the color cues could well have been the only cues attended to. Then there was the further fact that "random pattern" forms were generally irregular in contour, and there was a question as to whether they were more or less distinctive than geometric forms. Therefore, Warren (1953b) chose five forms of each variety (Fig. 15, top), cut them out in six different sizes, centered them on white cards that were 3 inches square, and paired cards with forms of equal size. Warren's rhesus monkeys did their best with pairs of cards whose forms belonged to different categories, but their learning rates for paired irregular forms equaled those for regular forms. Warren also found that, though the forms were paired for size, a positive relationship existed between performance levels and the sizes of the forms regardless of the nature of the forms (Fig. 15, bottom).

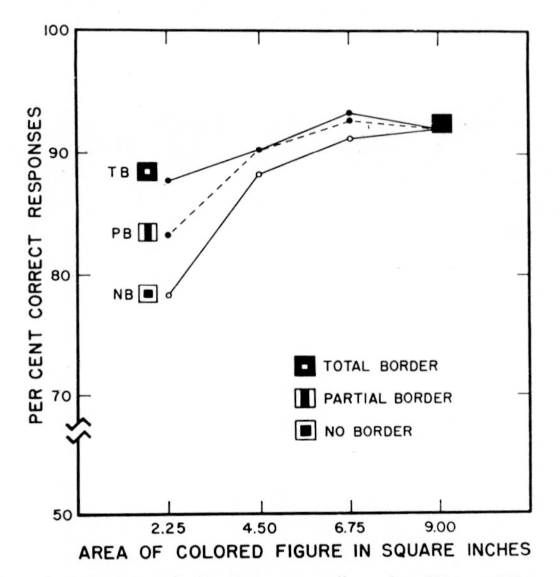

FIG. 16. Results of Warren's "color-pattern" study. (From Warren, 1953a.)

In an earlier experiment with this same group of monkeys, Warren (1953a) had observed that size effects are markedly a function of the loci of the cues that one displays upon a pair of plaques. Here he worked with differential colors that appeared within the centers of cards or upon their borders, or were placed within a strip, centered left and right, that ran between the near and far edges. Performance was related to cue area, but size effects were greatest when the colored areas were centered on the cards and thus were surrounded by white borders (Fig. 16).

In a third experiment, Warren (1953c) worked with cards on which the cues were centrally displayed and asked, again, if differences in color, form, and size have additive effects upon performance. Here he found that problems that presented color cues were solved just as rapidly as those in which the centered stimuli were different with respect to form and color, color and size, or color, form, and size combined. Form alone, or size alone, or form plus size yielded lower levels of performance, and there was no clear-cut proof of additivity of differences in size and form (Fig. 17). Color, as in Harlow's (1945d) and Meyer and Harlow's (1949) works, not only was the most effective cue: if present, it appeared to be the one effective cue as far as rhesus monkeys were concerned.

Such a view was nonetheless invalidated by a study in which Warren (1954) demonstrated that subtraction of a cue can bring about a change when the addition of that cue does not. Here he worked with large and small colored plywood forms, and his general method was to pair them

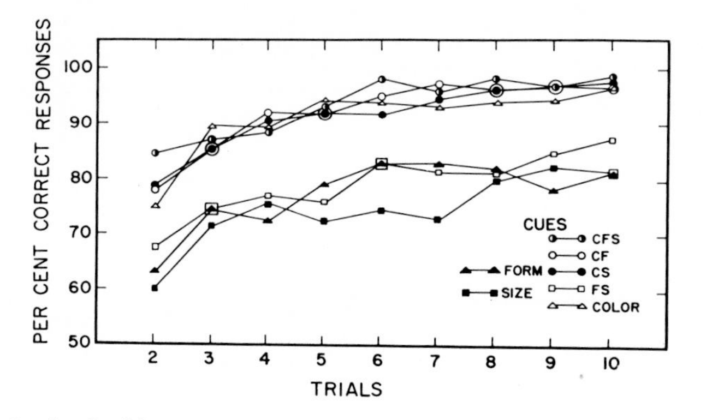

FIG. 17. Lack of additivity of color, form, and size. The curves are mean within-problem learning curves from which first-trial scores have been omitted. (From Warren, 1953c.)

so that they would differ with respect to color, size, and form when first presented to the rhesus monkeys. Then, when several training trials had been given, he reduced the number of such cues to one or two, and he found that this would always yield at least a mild disturbance of the animal's performance (Fig. 18). Even so, when differences in color were maintained, the added maintenance of form or size differences had no effect upon performance levels in the trials that followed the reduction of the cues. Therefore, one can show that groups of cues will fail to add when one can prove, by independent methods, that the monkey has, in fact, attended to them all when they are simultaneously present.

It appears that balance in the potencies of cues is crucial to occurrence of summation. Thus, in the experiment just described, form and size were reasonably matched in terms of post-reduction learning rates, and performance was better with combined differences in form and size than with differences in either of these cues alone. Even color, Warren (1956) found, will sum with other cues, but notably the latter observation was obtained with colors that were not demonstrably superior to the forms that he employed. Hence the question, at this stage, was why it is that forms and sizes that experimenters choose are usually inferior, as discriminative cues, to colors that experimenters choose.

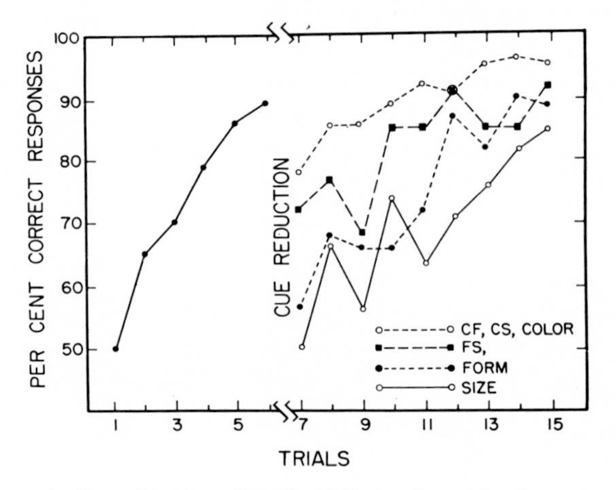

FIG. 18. Warren's data showing that the failure of a pair of cues to add does not imply that animals do not discriminate the presence of the less-effective cue. CF: color and form maintained. CS: color and size maintained. FS: form and size maintained. (From Warren, 1954.)

VI. THE RENAISSANCE OF CONTIGUITY

We shall next consider some experiments that deal with spatial stimulus-response relations. It will be recalled that Gellermann (1933a, 1933b) had found that it is hard to train a chimpanzee to make discriminations if the differential cues are not displayed directly at the points to which the subject makes responses. Later, Jenkins (1943) found a very similar effect with separations of a half a foot, and then there were no studies for another dozen years in what has since become an active field.

McClearn and Harlow (1954) were the first investigators to explore this problem parametrically. For their work, they built a special panel that contained a pair of narrow, vertical recesses (Fig. 19, top). Foodwells that were cut into the bottom of each recess were covered by two gray blocks; rhesus monkeys were required to move these blocks when making a discriminative choice. The cues were black and white blocks

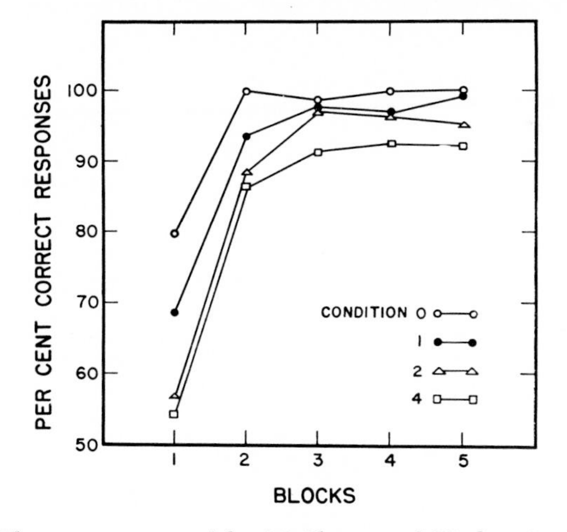

FIG. 19. Top: The apparatus used by McClearn and Harlow in their study of the consequences of remoteness of a pair of cues from the response (and, possibly, reward). The recess on the left shows an uncovered foodwell, and a covered one is shown on the right; the right recess contains a cue block that is placed remotely from the foodwell cover. Bottom: The results of the study, in 4-day blocks of trials. (From McClearn & Harlow, 1954.)

that were placed adjacent to, or 1, 2, or 4 inches over the two gray response blocks. McClearn and Harlow found that levels of performance were an inverse function of this distance, and that the discrepancy for the extreme conditions still was present after 20 days (Fig. 19, bottom).

A problem that was not resolved was whether this was due to cue-response or cue-reward relations. Jarvik (1953), who had recently reported one of two experiments devoted to the problem, emphasized the probable importance of a spatial cue-reward proximity effect. He had worked with bits of bread that served as both the cues and the rewards for learning of discriminations by chimpanzees, rhesus monkeys, and spider monkeys (*Ateles*). The bread was colored red or green with vegetable dyes, and some pieces were of natural flavor; others were sweetened or were made distasteful by immersion in a mixture of red pepper, quinine, and bile. Jarvik found that chimpanzees and monkeys quickly learn to pick the sweetened or unflavored bread, but he found that monkeys do not transfer such discriminations to a pair of colored plaques. Transfer was obtained when colored celluloid was pasted to or placed upon uncolored bread, but not when colored celluloid was laid adjacent to uncolored bread or over two foodwells containing uncolored bread.

Jarvik (1956), in his second study, worked with peanut shells, only some of which contained a nut. The shells were dyed with different colors so that they could serve as cues for a discriminative choice. Jarvik also used a special set of formboard plaques that were colored on their tops and bottoms, and which each contained a foodwell on its underside into which a peanut could be fitted and secured through the use of transparent tape. Jarvik found that chimpanzees can readily acquire discriminations in these situations, and that standard formboard methods are inferior to the use of plaques that have the foodwells cut into them.

The next experiment to emphasize the role of spatial factors in discrimination was performed by Murphy and Miller (1955) in an object-discrimination-learning context. These investigators used a Klüver formboard to which they had added a platform. Objects thus could either be displayed upon the tray or placed 6 inches above it on the platform, and the subjects could be asked to push aside either a stimulus object or a distant, neutral foodwell cover. Monkeys that were trained to make responses to the cues quickly formed an object-discrimination-learning set (see Chapter 2 by Miles), but a group that made remote responses failed to learn when trained with nearly 700 three-trial problems (Fig. 20). Further, when the monkeys that had formed a learning set were switched to discontiguous responding, their performance level fell until it was about the same as that of monkeys that had not.

In a second study, Murphy and Miller (1958) tried to see if these effects had been produced by cue-response, cue-reward, response-reward, or cue-response-reward separations. In this study, they employed a double Klüver formboard with one tray 7 inches above the other. Rhesus monkeys were the subjects, and conditions ranged from customary presentations with cue, response, and reward in the same place to a situation where the cues were in one place, the animal responded somewhere else, and food was then delivered in a distant foodwell when the choice response had been correct. Learning of a black-white discrimination habit was most rapid when the monkeys made their choice responses to the cues. Given this, the spatial contiguity between the cue and the reward was unimportant. Notably, one subject that performed successfully with all variations of the task did so only when it was observed to touch a cue before it moved a neutral response block.

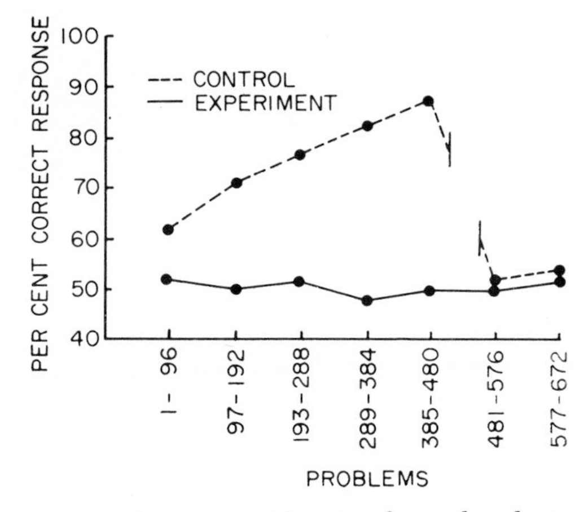

Fig. 20. Learning-set performance with pairs of stimulus objects displayed 6 inches above a test tray (Experimental) or on the tray (Control). The measure is percentage of correct responses on trials 2 and 3 of each problem, and the control group was shifted to the experimental condition after 480 problems. (After Murphy & Miller, 1955.)

One could view these studies as a much-belated proof that Spence (1937) had been essentially correct in his conclusions as to why some apparatuses are so much more efficient than are others. Murphy and Miller's studies, in particular, had shown that problems such as Bierens de Haan's are less a function of what primates can perceive than of the way the cues have been presented. The power of the formboard method thus appeared to be, in no small way, dependent on the fact that spatial contiguities as well as those of time are potent factors in the learning process.

VII. ON CONTIGUITY AND COLOR PATTERNS

We should next recall that Warren (1953a) found that rates of learning vary with the size of cues, and that this is markedly the case for colored squares displayed within the centers of two plaques. It was thought that this effect could well be transient, but Blazek and Harlow (1955) next observed that differences in difficulty thus produced are not reducible by interproblem practice (Fig. 21).

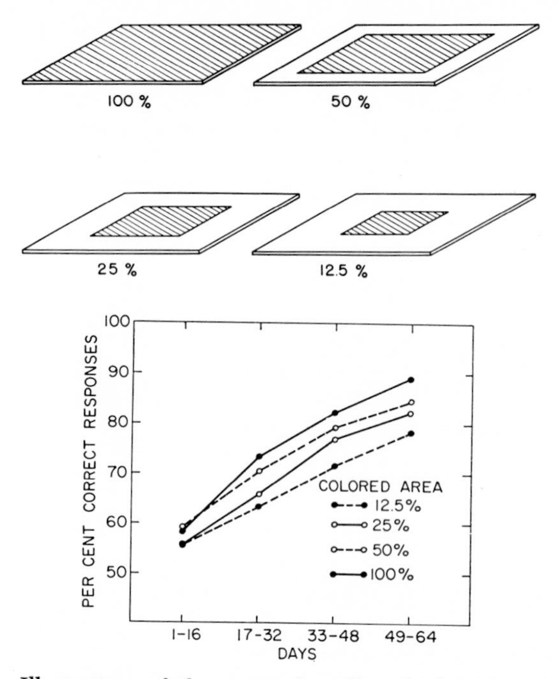

FIG. 21. Top: Illustrations of the types of cardboard plaque used by Blazek and Harlow. Cards used in a given pair differed in color, but not in size of cue. Bottom: The learning-set functions that resulted. The days of practice represent 768 six-trial discrimination problems. (From Blazek & Harlow, 1955.)

This result, in principle, was duplicated by the outcome of a study by Schrier and Harlow (1956) that examined the effects of size of center cues as well as amount of reward. The subjects in this study were cynomolgus monkeys (*Macaca irus*), and their curves for learning-set formation turned out to be almost absolutely parallel when size of cue was the parameter (Fig. 22). This persistent size effect was independent of effects on performance brought about by different quantities of food reward (see Chapter 2 by Miles).

In a second study, Schrier and Harlow (1957) modified a test tray by gluing two white cards on its surface. These two cards had centered holes that were cut to match the foodwells. Hence, when two small colored chips covered the foodwells (Fig. 23C), the tray appeared identical to one upon which Warren center-cue color-pattern cards had been displayed (Fig. 23A). The difference was that rhesus monkeys, when they made their choice responses, pushed aside the colored central chips, and they thus manipulated differential cues rather than irrelevant white borders. Rates of learning-set formation with these part-card prob-

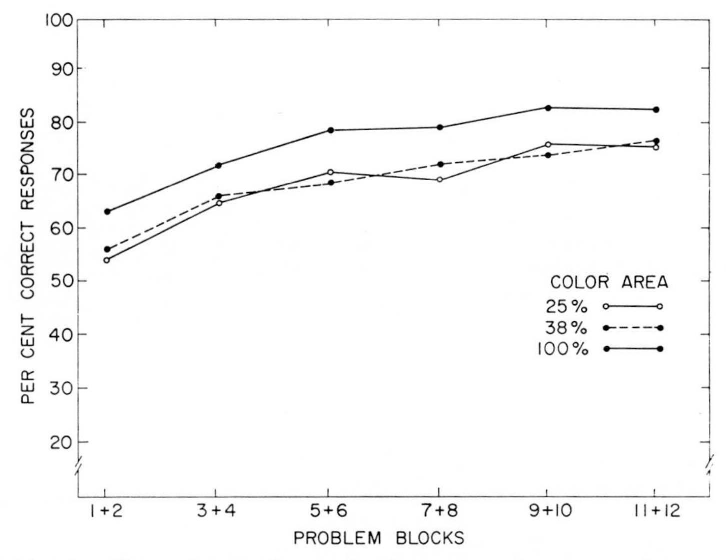

FIG. 22. The effects of area of central color cues on learning-set formation. Note that these three functions show no tendency to converge with interproblem practice. (After Schrier & Harlow, 1956.)

lems were compared with those obtained with whole cards, and with cards in which the entire 9 square inches of the surface were differentially colored. As expected, monkeys did quite poorly with the whole cards that presented central color cues, but they did much better when they moved aside a chip that lay on a firmly mounted background card (Fig. 24). In fact, they did about as well with little colored chips as with the wholly-colored pairs of cards, and practice with the part-card problems helped a group that then was trained with Warren whole-card problems. It was thus established that the center-cue effect of Warren, and of Blazek and Harlow, had to do with how the monkey made a choice response and not with size of color cue as such.

One could view this outcome as a kind of proof that spatial contiguity effects contribute to the problems that lower primates have with cues displayed upon two identical bases. Yet, this was a difficult conclusion to accept because of the dimensions it implied. The widest borders that had yet been used with 3-inch-square cards were merely of the order of an inch, and borders that were even narrower than this

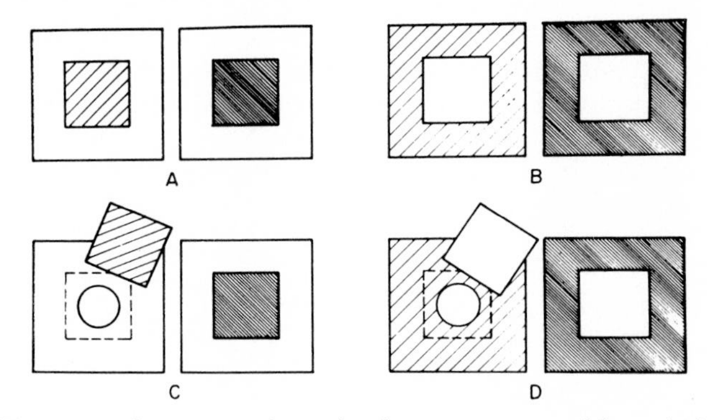

FIG. 23. Four spatial variants of a color-discrimination problem. (A) whole-card center-cue; (B) whole-card border-cue; (C) part-card center-cue; (D) part-card border-cue. C and D are shown with a foodwell uncovered after a response. (From Stollnitz, in press.)

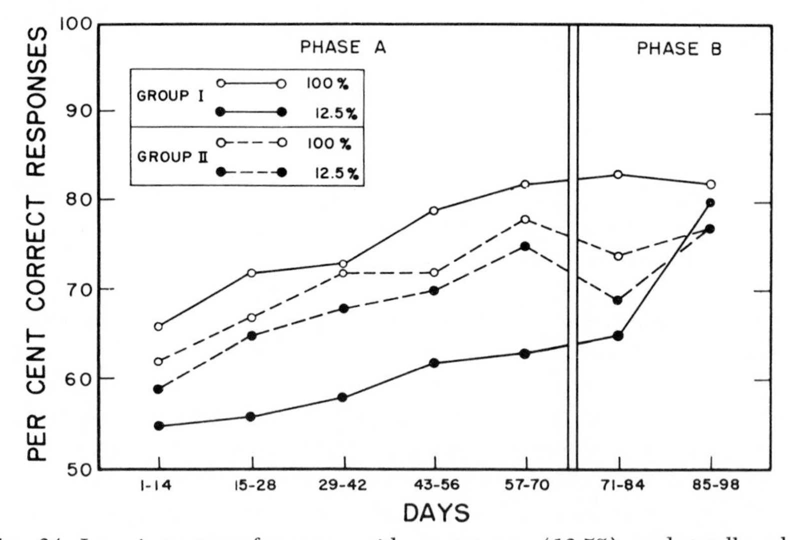

FIG. 24. Learning-set performance with center-cue (12.5%) and totally-colored (100%) cards. In Phase A, in 12.5% problems, group I worked with whole cards; group II pushed aside the colored cue-chips of part cards. In Phase B, conditions for the two groups were reversed, but otherwise conditions were the same. Note the marked rise in the performance of group I monkeys when they were trained to move the colored cue-chips; also note that group II monkeys still did well when they were shifted to the whole-card condition. (From Schrier & Harlow, 1957.)

had had some very durable effects. Thus, to say that spatial contiguity, alone, was at the bottom of the border problem was to say that miniscule remotenesses of cues can markedly reduce their potencies.

Such a view was nonetheless in keeping with results obtained in an experiment with rhesus and cynomolgus macaques that involved discrimination of two colored forms surrounded by concentric colored rings (Riopelle *et al.*, 1958). The enclosure of the colored forms by colored rings that were themselves irrelevant for learning had a very marked effect upon the learning rates for problems based upon the central cues. Further, when discriminative color cues appeared within the rings instead of in the centers, difficulty seemed to be a function of the distance of the cue-rings from the outer edges (Fig. 25).

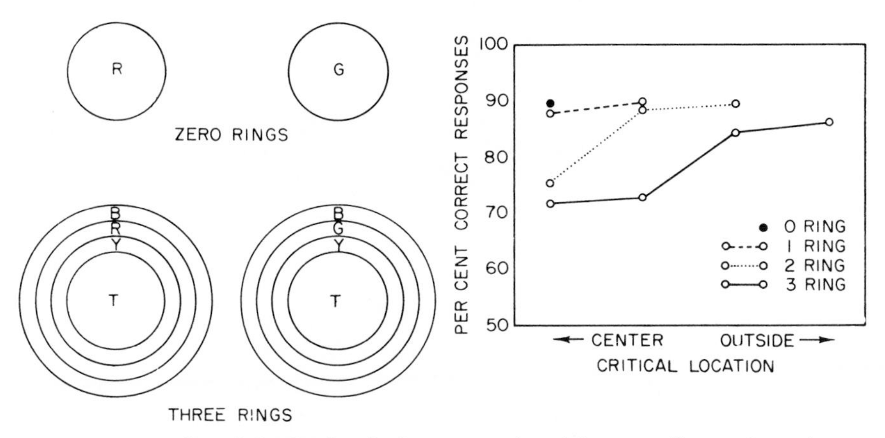

FIG. 25. Examples of the kinds of plaques employed by Riopelle *et al.*, and some of the results that they obtained. The central portion of a plaque and each ring was 2 square inches in area. Letters represent colors. (After Riopelle *et al.*, 1958.)

It could still be argued that these findings were, in part, a function of the handling of the cues; that is, that manipulation has a consequence apart from contiguity effects. Schuck (1960), however, failed to find support for such a view when he used Schrier and Harlow's (1957) part-card method to examine cases in which color cues are placed upon the borders (Fig. 23, B and D) rather than the centers (Fig. 23, A and C). Shuck confirmed the previous results with center cues, and found that, for the animals that learned, border-cue performance levels were as high when his rhesus monkeys moved the central chips alone as when they moved the comparable whole cards (Fig. 26). Hence, manipulation of irrelevant white chips did not suppress discrimination learning, and the contiguity relationship was all that seemed to matter in these situations.

Schuck went on to demonstrate that changes in response can have some other notable effects. To do this, he designed cards with strips of color cue along one edge. In his main experiment, his monkeys had been trained to push the cards away from them; this response had been enforced by narrow metal guides and pins that held the two discriminanda. Now, however, half the group was trained to pull the cards, and all the monkeys then were given practice with discrimination problems in which colored strips were on the nearest and the farthest edges.

As expected, animals that pushed the cards away mastered near-cue problems very quickly, and they also did extremely poorly when the cues were on the distant edges of the cards. But the striking finding was

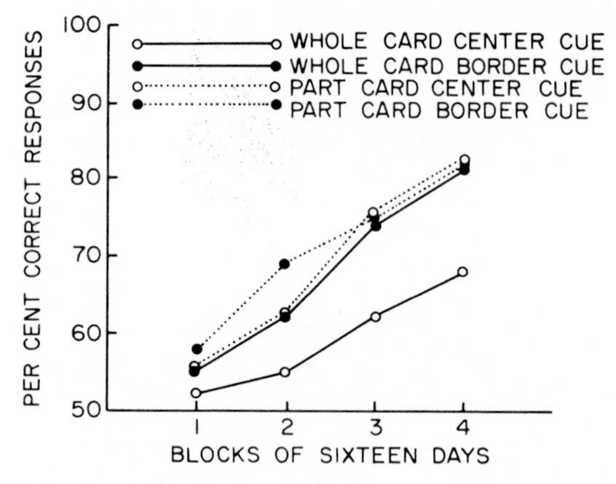

FIG. 26. Learning-set performance with four spatial variants of color-pattern-discrimination problems used by Schuck (1960).

that opposite results were not obtained when monkeys pulled the cards; rather, they did just about as badly with the far- as with the near-cue stimulus arrangement. Schuck, however, noted that the monkeys of this group often touched the nearest edges first, and he held that the persistence of the old response had led to a suppression of performance.

But, despite the complication of this last result, it was clear that properties of cues are more a function of response location than they are of anything intrinsic to the cues. Monkeys seem to sample from the point of their response, and stimuli an inch or so away may simply not be noticed by them even though they seem quite obvious to the experimenter. For example, pairs of forms with different apices may actually appear to be the same to monkeys if their bases are identical and subjects touch the bases when they make responses. Hence, not only base effects, but dominance effects and additivities of groups of cues are markedly dependent on the sampling biases identified in these experiments.

VIII. ON CONTIGUITY AND AUTOMATION

We shall next consider what this rather startling fact should mean to anyone who plans to build an automatic two-choice training apparatus for research with nonhuman primates. It is rather obviously not a good idea to separate responses from the cues, but this is exactly how a number of the first devices of this sort were put together. Typically, the stimuli to be discriminated were presented in a pair of windows, and the subject was required to signify a choice by pressing on a key below one window (Fig. 27). Such arrangements have a very strong relationship to Bierens de Haan's (1927) apparatus (Fig. 3), and in this respect they represent a backward step so far as training methods are concerned.

FIG. 27. An automatic primate-training apparatus in which the discriminanda, manipulanda, and food cup are all separated from each other. (From Battig *et al.*, 1960, who used this apparatus for delayed-alternation training.)

Certain problems, nonetheless, can hardly be approached unless the choice responses are remote, and we have searched for ways of overcoming the effects that render such devices inefficient. The first of our arrangements was constructed with the hope that S-R contiguity effects would disappear if presentations of the stimuli were under the control of a response. In this apparatus (McConnell *et al.*, 1959), monkeys were required to put their heads into a plastic mask; this response was followed by the rear-projection of a pair of stimuli upon two screens (Fig. 28). The latter, which were placed at the level of the mask, were of the size of standard Warren cards and were mounted in a forward wall that also held a pair of corresponding pellet chutes.

FIG. 28. The animal compartment of the first Ohio State Apparatus. The monkey is making a contiguous response after having made a mask response. The plan of the dividing wall is shown at right; note the mask, the sloping arm-holes, and the photo relay that detected mask responses. The dimensions of the box are given by the marks, which are spaced 1 inch apart. (From Meyer *et al.*, 1961.)

In the first experiments performed with this device (Meyer *et al.*, 1961), rhesus monkeys were required to learn a discrimination based upon a difference in brightness. Initially, the animals were trained to touch the chutes while sitting on a large metal plate; weak currents passing through their bodies were employed to activate pellet-dispenser circuitry (cf. Klüver, 1935). Subsequently, prior to introduction of the panel that contained the orienting face mask, the animals were trained to touch the pellet chute beneath a suddenly illuminated screen. This was followed by a briefer set of training trials in which the food appeared within the chutes, but subjects were required to make contiguous responses to the overlying screens.

Naive monkeys failed to learn the first of these two tasks within the course of several hundred trials, but they all did well when they were next required to make contiguous responses to the screens. Their discriminations then broke down when they were forced to touch the two delivery chutes again, and differences related to location of response were found in eight successive alternations (Fig. 29).

At this point, the panel with the mask was introduced, and thereafter presentations of the stimuli were under the control of the monkey. The forward pellet chutes were both removed, and sugar-pellet reinforcers were delivered to a cup that was placed within the mask. The monkeys made contiguous responses, and measurements of levels of performance were obtained as varying delays were introduced between the presentation of the stimulus and activation of manipulanda. It was found

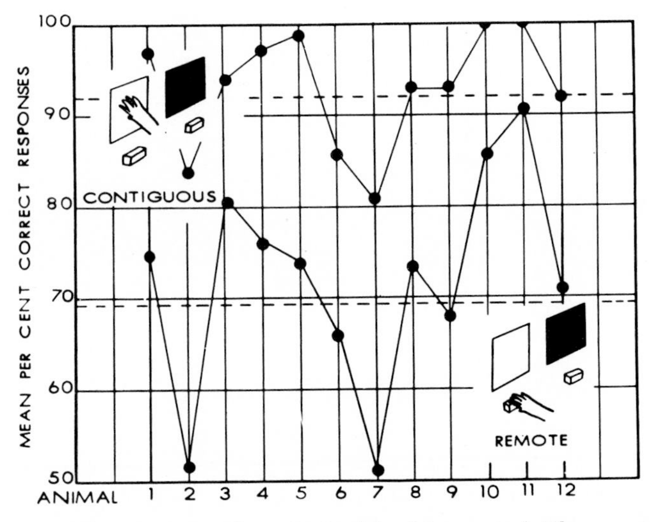

FIG. 29. The S-R contiguity effect obtained by Meyer *et al.* The remote responses were made to pellet chutes 2 inches beneath the screens. (From Meyer *et al.*, 1961.)

that, for this task, the optimal delay was somewhere close to 0.5 second, and except where specified, this delay was used in all the rest of our investigations.

A subsequent experiment by Schuck *et al.* (1961), conducted with this same group of monkeys, asked if the response-contingent-stimulus approach had overcome the spatial sampling bias. In this work, the animals were forced to make responses to a pair of small, black buttons that were mounted squarely in the centers of the screens. Center-cue and border-cue color-pattern problems were employed, and the monkeys, as before, obtained a sugar pellet from a lip cup when they were correct. Here, in contrast with the outcome commonly obtained with Warren cards presented with the formboard, problems of the center-cue variety were learned more rapidly than border-cue problems (Fig. 30). Clearly, then, the locus of the monkey's choice response continued to determine what it saw; spatial sampling biases had not been mitigated by the introduction of the mask.

At this point, it seemed that there was no alternative to having monkeys push upon displays, and that even this would not suffice unless the screens that one employed were reasonably small. Therefore, Horel *et al.* (1961) looked into the feasibility of using patterns that would be so small that they would be covered over by the monkey's hand when it made its choice responses to them. Only one discrimination problem was employed, but it was extremely difficult: the stimuli were two

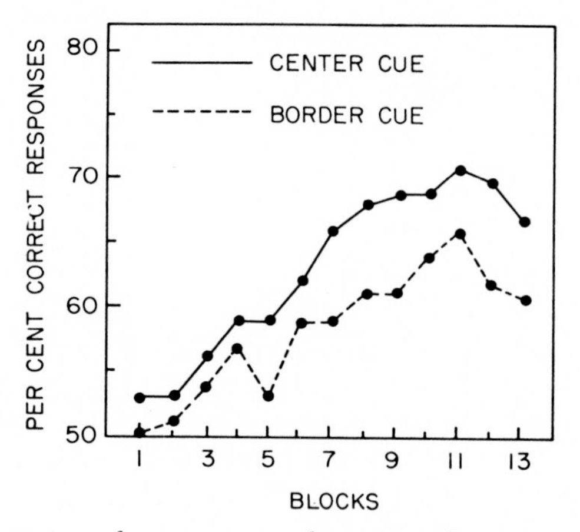

FIG. 30. Learning-set performance on color-pattern-discrimination problems presented in the first Ohio State Apparatus. Center cues were 2 square inches in area, but nonetheless were found to be superior to border cues of more than three times that area. (After Schuck *et al.*, 1961.)

small discs that each contained a bar that ran from side to side or up and down. These appeared upon two screens similar to those that Schuck and his associates had used, and they were displayed above, below, between, or outside two small buttons centered in the screens.

It was found that placement of the discs beneath the buttons, where the monkey's hands would cover them, was similar in its effects upon performance to projection of the discs above the buttons. This result suggested that the choices were controlled by sampling that preceded the response, and that the designer of displays could thus ignore what happens at the moment of response. Therefore, at this point, the proper strategy appeared to be to concentrate upon the study of the consequences of manipulation of the cues *before* the choice response occurred.

A subsequent development, however, changed our mind, for, in some work by Sheridan *et al.* (1962), we found that some cue changes occurring *after* the response can markedly affect the learning process. In this study, it was noted that the response-contingent disappearance of a pair of cues has no effect upon the levels found when both the cues persist beyond the time of the response. However, individual persistences of cues, whether positive or negative, yielded a substantial, and a reasonably stable, increment in accuracy of choice (Fig. 31). This result was interesting because it could not be interpreted in terms of any notion with respect to the effects of stimulus persistence upon strengths of single stimuli. Therefore, we decided to continue with our search for

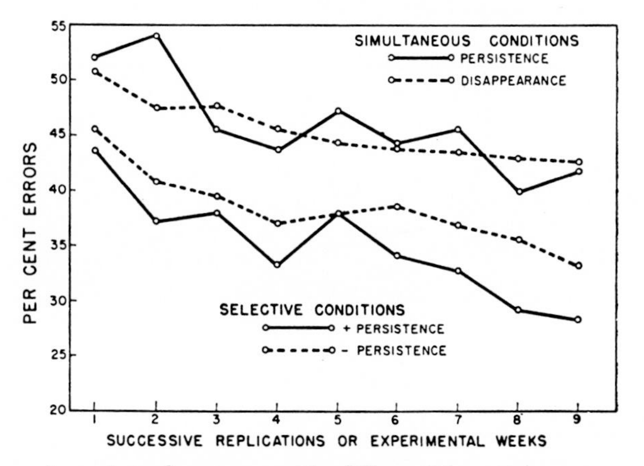

FIG. 31. Learning-set performance with differential postresponse persistences of cues in color-discrimination problems. (After Sheridan *et al.*, 1962.)

some arrangement of the test compartment that would make it possible for monkeys to perform with S-R discontiguous displays.

We next considered that the monkeys might attend to cues if they could not attend to their responses, and this possibility was tested in a study carried out by Otteson *et al.* (1962). The monkeys, once again, were asked to signify a choice by touching one of two pellet chutes. Under one condition, the stimuli and chutes were all within the monkey's field of view, and levels of performance under these circumstances were compared with the efficiency of learning when an opaque panel was inserted to divide the stimulus from the response compartment (Fig. 32). The animals were trained with problems generated by the pairing of black English letters and symbols centered on uniform white backgrounds. The monkeys did a little better when delivery chutes were hidden underneath the opaque panel, and it seemed that this approach could fruitfully be used in the construction of a new machine.

There were several other features that appeared to be desirable in such an apparatus. One of these was some device by which the animal could be prevented from performing choice responses during intervals of enforced observation. We had also found that there is simply no such thing as a first-class rear-projection screen, and we therefore wanted to be able to employ direct-projection screens in the display.

Accordingly, the new display consisted of a pair of square, shallow, horizontal foodwells (Fig. 33). These, like standard Warren cards, were 3 inches square, and stimuli appeared upon their floors. The animal could see both foodwells through a window, and could gain eventual access to them through two elbow-level holes that normally were covered by a single sliding panel. The monkey, in responding, had to move the

FIG. 32. The first Ohio State Apparatus as modified by Otteson *et al.* Note the panel that was used to segregate the stimuli from the manipulanda. (From Otteson *et al.*, 1962.)

slide to make a match between one arm-hole and an aperture within the panel it manipulated. The slide was locked, however, when the animal was not appropriately positioned at the window, and it also could be locked to move in just one way and thus enforce correctional responses.

Basically the new display was the inversion of the WGTA situation. The reward, instead of being covered by a plaque, was dropped upon the surface of a plaque. Hence, although a bias in the place of its appearance could not be completely done away with, the reward could land at any point within the plaque as well as showing up along an edge. It was thought that if the cue-reward relationship has some effect upon the sampling process, variations of this kind would counteract the bias and produce more uniform attending.

One of the most difficult of problems that we faced in the design of this new apparatus was the humdrum issue as to what could be employed for rewarding correct responses. We had used small sugar pellets in the first machine, but the monkeys did not seem to like them, and we therefore wished to find a pellet that would have a somewhat higher preferential value. After some experiments (Treichler *et al.*, 1962), we settled on the use of ordinary seed soybeans; these when roasted, are accepted by the animals, and do not often jam in magazines.

The apparatus thus completed was examined in a study executed by LeVere (1962). He began by training rhesus monkeys in its use, but intermixed this shaping with some practice with a set of Schuck-type color-pattern problems in a standard WGTA. Cues appeared as colored strips in any one of four positions on the edges of the cards; they were on the inside or the outside edges, or were near or far from the subjects.

FIG. 33. The second Ohio State Apparatus. Upper left: the mechanism of the sliding panel, showing one foodwell beyond the arm-hole in the center of the panel and a corner of the other foodwell beyond one of the slots by which the monkey moved the panel. Upper right: the monkey's box, showing the elbow-level ports and the window. Lower left: a monkey at work. Lower right: the box, the framework, the projector, and the group of mechanisms that controlled the program.

The expected biases were found, with near-cue problems being fairly easy, but there was an unexpected difference between the inside- and the outside-cue positions. Performances with outside cues fell between extremes obtained with near- and far-cue arrangements, but the inside-cue condition was about as poor as that in which the colored strips were far (Fig. 34).

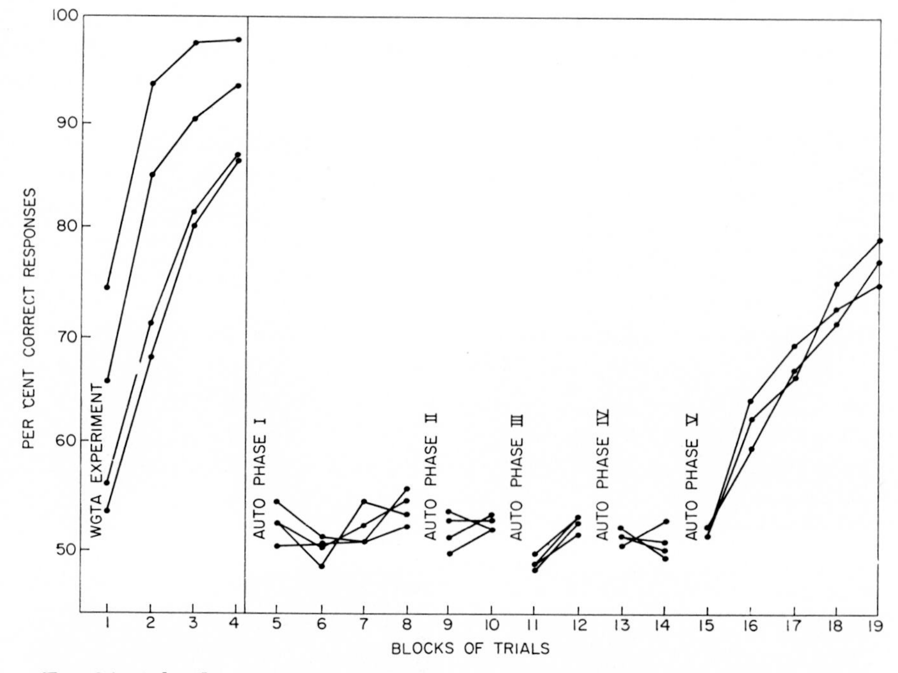

FIG. 34. Color-discrimination performance in the study by LeVere (1962). The learning-set curves for the WGTA experiment are those for near cues, outside cues, inside cues, and far cues in descending order of performance. Performance with near cues was already asymptotic after just three blocks of presentations, but ten such blocks of presentations (Phases I through IV) in the second automatic apparatus revealed no sign of learning. The monkeys later learned three discriminations when the foodwells were completely filled with color (Phase V).

The animals were subsequently given training with these problems in the new apparatus. By this time, they all had been conditioned to perform the necessary acts of operation, and had had experience in picking up the beans when these had been delivered to the foodwells. Since it was expected that the new machine would be superior to the WGTA, the same cue pairings were presented once again with their reward contingencies reversed.

In Phase I, the stimuli appeared within the foodwells when the monkey looked into the window, and the sliding panel was unlocked when this response had been maintained for one full second. If the monkey moved the panel so that it could reach into the proper foodwell at this time, the response of opening this port produced a bean that it could then pick up, remove, and eat. If the first response that it performed was incorrect, the animal was forced to make corrections, but the panel could be moved if, and only if, the monkey kept its head within the window.

There were several variations in conditions as the study of the new machine progressed. At one point, the observational delay was shortened to half a second (Phase II). At another, presentations of the pairs of cues were made independent of the monkey (Phase III). A third change consisted of the installation of a pair of buttons next to the foodwells; the monkey then was asked to press upon this hidden button after it had opened up a port (Phase IV). But, despite these variations, five rhesus monkeys wholly failed to learn these simple tasks, and they failed to do so after five times the practice that had been sufficient when these same animals were tested in the WGTA (Fig. 34).

These monkeys next were given training with a set of three concurrent color-discrimination problems (Phase V), and they all learned reasonably quickly when the wells were flooded with the differential cues. This result suggested that the prior difficulty could have been a function of the monkeys' sampling only from those portions of the apparatus where the soybeans were delivered. But this left the question as to why a sampling bias of this sort should have a consequence that is even worse for learning than the bias found when animals manipulate a card.

At any rate, it seemed to us appropriate to try an altogether different approach, and so we built a final apparatus to explore a possibility that still remained. In it, pairs of stimuli were once again returned to a position opposite a mask. They appeared upon two screens that lay above a pair of platforms to which animals gained access by depressing one of two doors that were interlocked so that one, and only one, would open. The occurrence of a choice also turned on lights, and the monkey could then see the platforms through two prisms that were so positioned that this view was mixed with what was shown upon the screens. Hence, a raisin placed upon a platform could appear to be at any point within the pattern that was being shown upon the rear-projection screen that lay directly over either platform (Fig. 35).

A group of six monkeys served as subjects for a study, carried out by Frank (1963), in which the five that learned to operate the new device were first given training with black-and-white pattern-discrimination problems obtained by pairing 12 patterns randomly selected from a representative collection of the ones presented in a book by Hornung (1946). Only one monkey did particularly well until correction methods were employed, and the rates were even then no better than they are for pattern-discrimination learning in the formboard context. The five monkeys then were trained on color-pattern-discrimination problems in

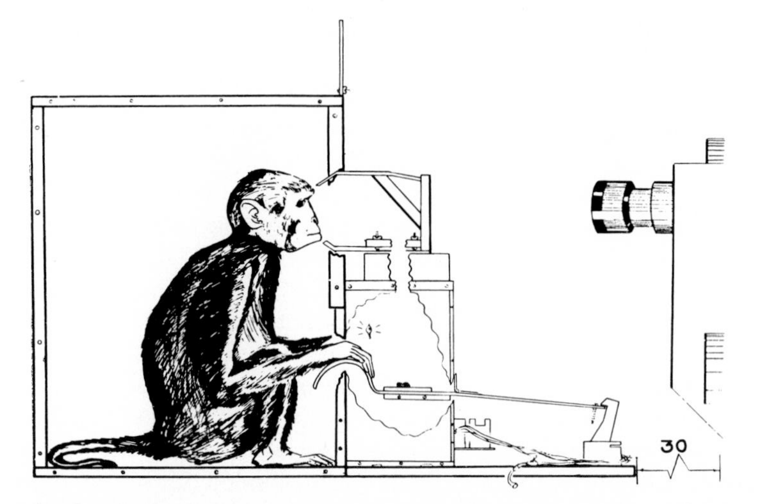

FIG. 35. The third Ohio State Apparatus. Here the monkey makes a discontiguous response, but the image of the food then appears together with the cues in the windows.

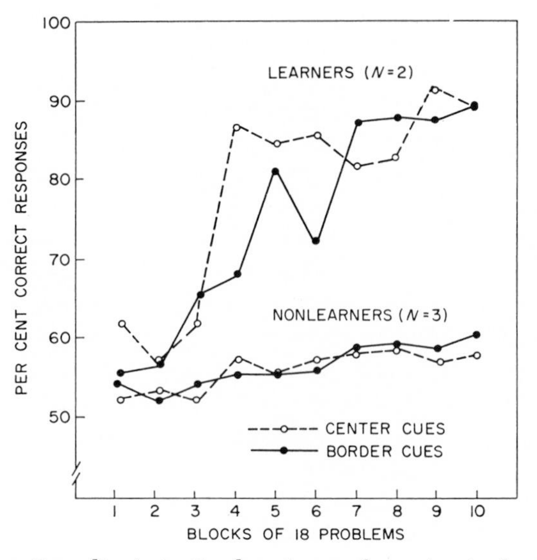

FIG. 36. Color-pattern-discrimination-learning-set formation in the third Ohio State Apparatus. (After Frank, 1963.)

which color cues occupied half of the 4 square inches of each screen. Levels of performance were about the same with center-cue and border-cue arrangements, but three monkeys of the group of five were still doing poorly after they had been exposed to 360 ten-trial problems (Fig. 36).

IX. SOME FINAL OBSERVATIONS AND CONCLUSIONS

It is clear that we have not, to date, discovered any S-R discontigous arrangement with which monkeys will perform as well as they perform when trained with manual WGTAs. Nor, so far as we have learned, has anybody else produced an apparatus of this class for which measured levels of efficiency for learning have been much more promising than ours. There are automatic apparatuses that work, but all of them involve displays in which the monkeys are required to make contiguous responses to discriminative stimuli.

The Wisconsin Automatic Test Apparatus (WATA) is an excellent example. This machine, designed by Polidora and Main (1963) is a

FIG. 37. The forward wall of the animal compartment of the WATA. The monkey views the wall and makes responses to the screens or to the underlying trays through a mask partition similar to that used in the first Ohio State Apparatus. The center screen was not used by Polidora and Fletcher (1964). (Photograph courtesy V. J. Polidora.)

punched-card-programmed apparatus. The initial studies (Polidora & Fletcher, 1964) carried out with the machine deserve examination in detail, for some points that have been made have several consequences in the field of methodology.

Basically, the studies were concerned with the effects of two modes of responding to a pair of visual cues that were presented in two windows when the subject put its head into a mask. Rhesus monkeys of one group were trained, initially, to make responses to the stimuli; monkeys of

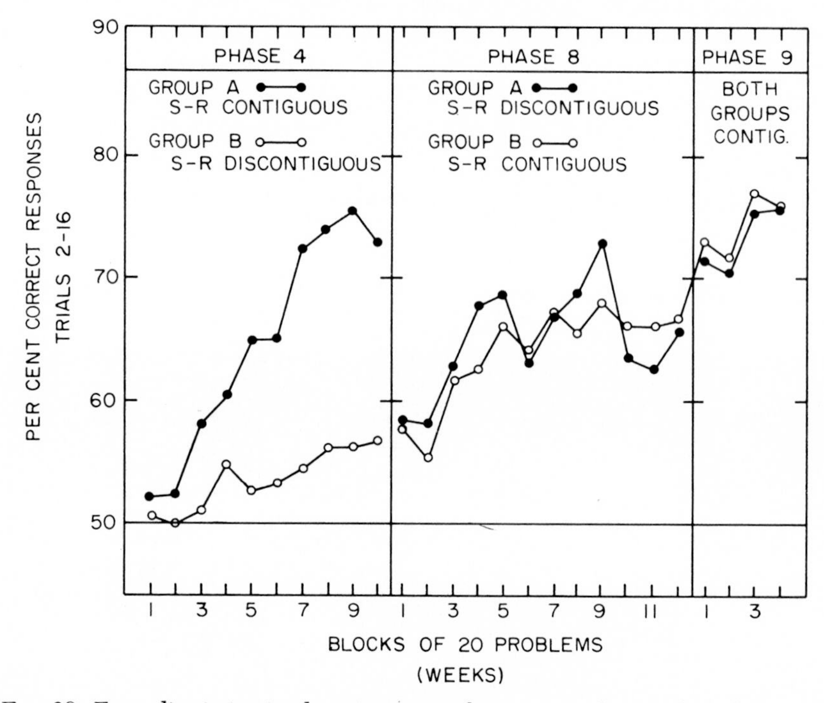

FIG. 38. Form-discrimination-learning-set performance with stimuli and responses spatially contiguous or discontiguous. (From Polidora & Fletcher, 1964.)

another group were trained to open trays that lay directly underneath the windows. The trays were also used for the delivery of rewards under both of the above conditions, and were kept locked open if the monkeys were required to choose by pressing on a stimulus (Fig. 37).

The animals were first trained to a high criterion on three preliminary problems. They were next presented with a set of 200 sixteen-trial form-discrimination problems. Next, the entire program was repeated with a switch in the response requirements for the groups, i.e., monkeys that had pressed the panels opened trays, and tray-pullers pressed the window panels. At this time, the monkeys were observed while they were working, and a daily protocol was kept of how each individual subject had reacted to the alteration in response requirements.

It was found that, during the first form-discrimination series, animals that pressed the panels that contained the cues did better than the tray-responders (Fig. 38, Phase 4). When response requirements for the two groups were switched, the functions for the two groups came together (Fig. 38, Phase 8). However, it was noted that this over-all outcome masked some differences within the groups. Thus the panel-pressers

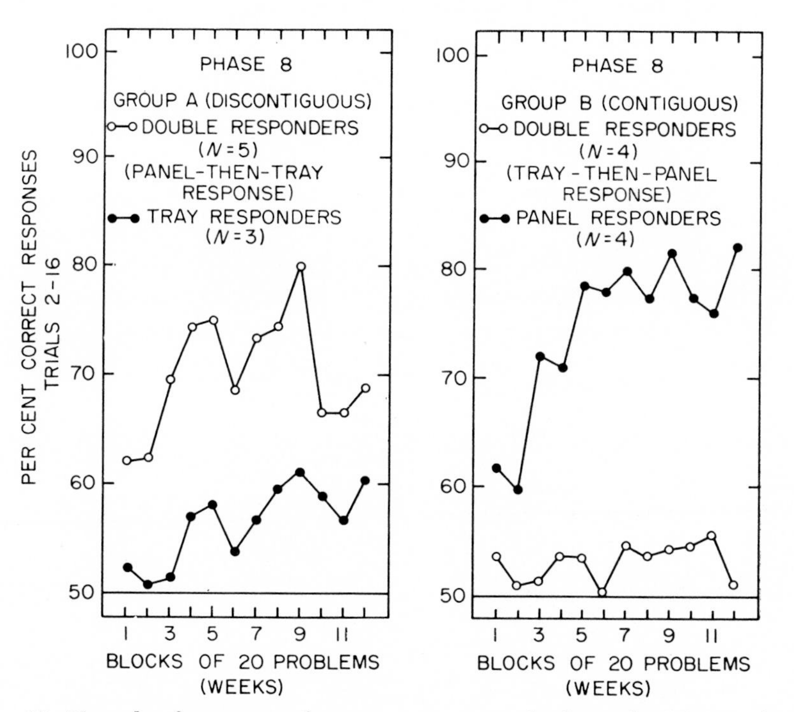

FIG. 39. The role of sequence of responses to manipulanda in the WATA. (From Polidora & Fletcher, 1964.)

that, when shifted to the trays, continued to respond to the panels first, did much better than the ones that lost this first response and went directly to the trays (Fig. 39, left). Similarly, animals within the other group that touched the trays before they touched the panels did no better now than they had done when they were not required to make contiguous responses (Fig. 39, right). Therefore, in this situation, monkeys seemed to sample from the locus of the first response, and did not so readily discriminate the cues associated with the next response.

This conclusion is important in that it suggests some things about the placement of food-dispensers. If a monkey, as it looks away from a cue locus, is past the point of sampling for that trial, then it should not matter where one puts a dispenser in relation to the stimulus display. One, in fact, should be enough, and we may thus regard the format of the Discrimination Apparatus for Discrete Trial Analysis (Pribram *et al.*, 1962) as being sound although it uses 16 different screens arranged around a single pellet chute (Fig. 40). We should note, however, that the principle of sampling from the locus of the first response implies that there can be some sampling complications when devices of this kind are utilized for work with problems that involve sequential tasks (Pribram, 1963).

The first-response principle, so far as it applies to shorter chains, appears to be in keeping with results that Wunderlich and Dorff (1965) have very recently obtained with rhesus monkeys in the WGTA. These investigators made transparent plaques that would hold a set of stan-

FIG. 40. The forward wall of the animal compartment of the Discrimination Apparatus for Discrete Trial Analysis (DADTA). (From Pribram *et al.*, 1962.)

dard color-pattern cards, and studied all the possible combinations of stimulus-response and stimulus-reward spatial contiguity and separation (Fig. 41). They observed that, if a monkey was required to make contiguous responses to the cards, then it did not matter whether the reward was placed beneath the card (C-7 and C-8) or in a distant foodwell (C-5 and C-6). Further, the effects of center- versus border-cues were absent when responses were remote (C-1, C-2, C-3, and C-4); a reward, when placed beneath a card, did not itself produce the center-border sampling bias. We should note that this implies that bases can be used without producing classic base effects so long as one makes certain that the animal cannot displace a mounted object by the base.

It has long been wondered whether the efficiency of learning when two objects are employed reflects the fact that haptic cues support the visual cues that complex, solid stimuli provide. The only recent study of this possibility is Peterson and Rumbaugh's (1963), who observed that squirrel monkeys did no better when they touched an object than they did when they were asked to touch a transparent panel placed in front of the two objects (Fig. 42). Hence, there is no reason, at the moment, to believe that two-dimensional patterns have to be the poor discriminanda that experiments to date have almost always seemed to say they are.

Cue-response separation along one line of sight, as in Peterson and Rumbaugh's study, is the only circumstance in which it seems to make but very little difference whether monkeys actually make contact with

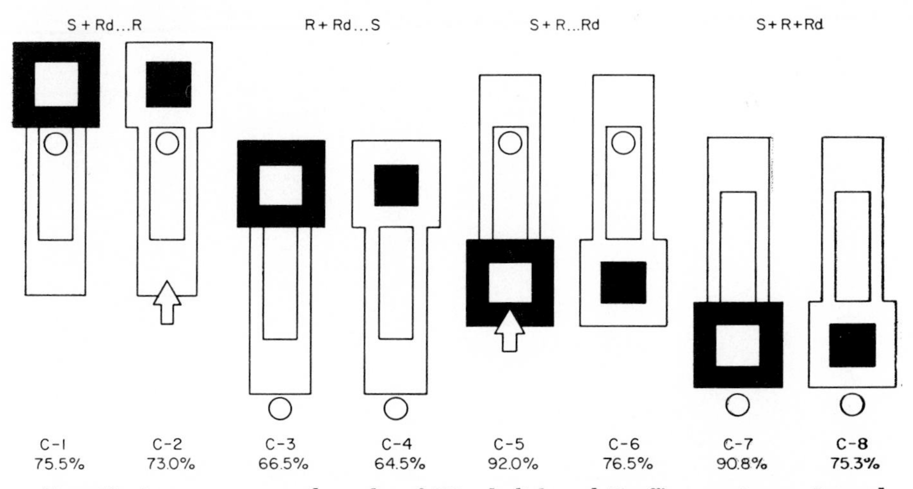

FIG. 41. Arrangements and results of Wunderlich and Dorff's experiment. In each condition, monkeys were required to push the double plaques in a direction that is upward in this diagram. Circles show where foodwells were uncovered when the plaques had been displaced. Percentage of correct responses is shown for each condition. (After Wunderlich & Dorff, 1965.)

the cues. Angular separations of the cues from the point of the response seem always to produce a sampling bias; physical arrangements for subverting its effects remain, if they exist, to be discovered. It is wholly possible that systematic shaping (Stollnitz & Schrier, 1962; Schrier *et al.*, 1963) is the only practical approach in situations where it is essential that discriminanda be separated from manipulanda.

Investigators who can use contiguous displays should make the cues as small as possible. We have never worked this way, but John S. Stamm has said, informally, that monkeys quickly learn discriminations in an apparatus he has built that seems to fit this notion fairly well (Fig. 43). This has disadvantages, but not one of them can match the disadvantages entailed by the persistent gradients that have been shown to have effects within a fraction of an inch.

Whether, even with small cues presented in this way, "random-pattern" learning rates will match the rates obtained with complex objects and the Klüver board is something that remains to be established. The problem is important, for if differences exist, the WGTA will have withstood another challenge to its place of leadership among available procedures in this field. Monkeys will continue to select tobacco tins, tomato cans,

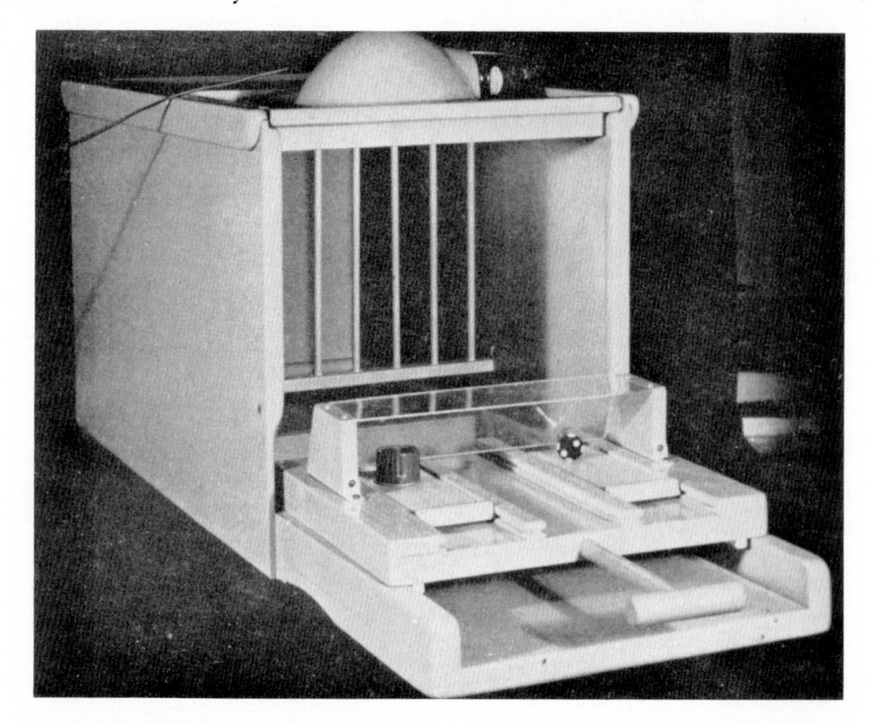

FIG. 42. WGTA-type apparatus for squirrel monkeys that was used with and without the plastic screen in front of the objects. (From Peterson & Rumbaugh, 1963.)

and other bits of trash, and they will acquire discriminations rapidly instead of taking many days or weeks. Though the manual method, in this day of the machine, is pitifully old-fashioned, it has set a standard of efficiency that hasn't been approached by any other apparatus yet.

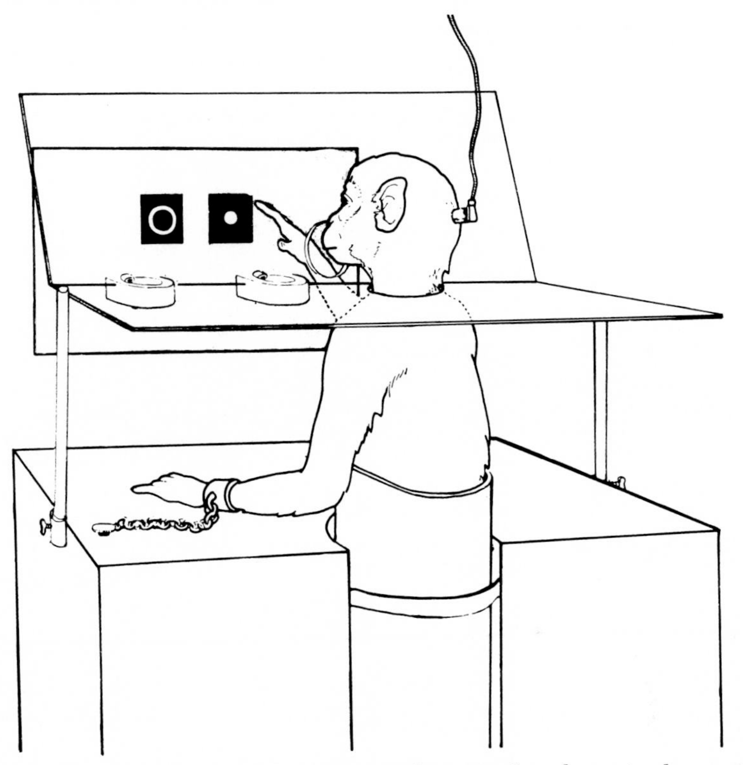

FIG. 43. The apparatus used by Stamm. The animal makes stimulus-contiguous responses to a pair of small discriminanda. (Drawing courtesy John S. Stamm.)

REFERENCES

Battig, K., Rosvold, H. E., & Mishkin, M. (1960). Comparison of the effects of frontal and caudate lesions on delayed response and alternation in monkeys. *J. comp. physiol. Psychol.* **53**, 400.

Berkson, G. (1962). Food motivation and delayed response in gibbons. *J. comp. physiol. Psychol.* **55**, 1040.

Bierens de Haan, J. A. (1927). Versuche über das Sehen der Affen. IV. Das Erkennen gleichförmiger und ungleichförmiger Gegenstände bei niederen Affen. V. Erkennen Affen in zweidimensionalen Abbildungen ihnen bekannte Gegenstände wieder? *Z. vergl. Physiol.* **5**, 699.

Blazek, Nancy C., & Harlow, H. F. (1955). Persistence of performance differences on discriminations of varying difficulty. *J. comp. physiol. Psychol.* **48**, 86.

Butler, R. A., & Harlow, H. F. (1956). The effects of auditory distraction on the performance of monkeys. *J. gen. Psychol.* **54**, 15.

Davis, R. T. (1956). Problem-solving behavior of monkeys as a function of work variables. *J. comp. physiol. Psychol.* **49**, 499.

Fay, J. C., Miller, J. D., & Harlow, H. F. (1953). Incentive size, food deprivation, and food preference. *J. comp. physiol. Psychol.* **46**, 13.

Frank, J. A. (1963). Effects of visual contiguity of cue, reward, and response on primate pattern discrimination in an automated apparatus. Unpublished master's thesis, The Ohio State University.

Gellermann, L. W. (1933a). Form discrimination in chimpanzees and two-year-old children: I. Form (triangularity) *per se*. *J. genet. Psychol.* **42**, 3.

Gellermann, L. W. (1933b). Form discrimination in chimpanzees and two-year-old children: II. Form versus background. *J. genet. Psychol.* **42**, 28.

Gross, C. G. (1963). Effect of deprivation on delayed response and delayed alternation performance by normal and brain operated monkeys. *J. comp. physiol. Psychol.* **56**, 48.

Harlow, H. F. (1942). Response by rhesus monkeys to stimuli having multiple-sign values. *In* "Studies in Personality" (Q. McNemar & Maud A. Merrill, eds.), pp. 105-123. McGraw-Hill, New York.

Harlow, H. F. (1944). Studies in discrimination learning by monkeys: I. The learning of discrimination series and the reversal of discrimination series. *J. gen. Psychol.* **30**, 3.

Harlow, H. F. (1945a). Studies in discrimination learning by monkeys: III. Factors influencing the facility of solution of discrimination problems by rhesus monkeys. *J. gen. Psychol.* **32**, 213.

Harlow, H. F. (1945b). Studies in discrimination learning by monkeys: IV. Relative difficulty of discriminations between stimulus-objects and between comparable patterns with homogeneous and with heterogeneous grounds. *J. gen. Psychol.* **32**, 317.

Harlow, H. F. (1945c). Studies in discrimination learning in monkeys: V. Initial performance by experimentally naive monkeys on stimulus-object and pattern discriminations. *J. gen. Psychol.* **33**, 3.

Harlow, H. F. (1945d). Studies in discrimination learning by monkeys: VI. Discriminations between stimuli differing in both color and form, only in color, and only in form. *J. gen. Psychol.* **33**, 225.

Harlow, H. F. (1949). The formation of learning sets. *Psychol. Rev.* **56**, 51.

Harlow, H. F. (1951). Primate learning. *In* "Comparative Psychology" (C. P. Stone, ed.), 3rd ed., pp. 183–238. Prentice-Hall, New York.

Harlow, H. F. (1959). Learning set and error factor theory. *In* "Psychology: A Study of a Science" (S. Koch, ed.), Vol. 2, pp. 492–537. McGraw-Hill, New York.

Harlow, H. F., & Bromer, J. A. (1938). A test-apparatus for monkeys. *Psychol. Rec.* **19**, 434.

Harlow, H. F., & Meyer, D. R. (1952). Paired-comparisons scales for monkey rewards. *J. comp. physiol. Psychol.* **45**, 73.

Harlow, H. F., & Warren, J. M. (1952). Formation and transfer of discrimination learning sets. *J. comp. physiol. Psychol.* **45**, 482.

Horel, J. A., Schuck, J. R., & Meyer, D. R. (1961). Effects of spatial stimulus arrangements upon discrimination learning by monkeys. *J. comp. physiol. Psychol.* **54**, 546.

Hornung, C. P. (1946). "Handbook of Designs and Devices." Dover, New York.

Jarvik, M. E. (1953). Discrimination of colored food and food signs by primates. *J. comp. physiol. Psychol.* **46**, 390.

Jarvik, M. E. (1956). Simple color discrimination in chimpanzees: Effect of varying contiguity between cue and incentive. *J. comp. physiol. Psychol.* **49,** 492.

Jenkins, W. O. (1943). A spatial factor in chimpanzee learning. *J. comp. Psychol.* **35,** 81.

Johnson, H. M. (1914). Visual pattern-discrimination in the vertebrates—I. Problems and methods. *J. anim. Behav.* **4,** 319.

Johnson, H. M. (1916). Visual pattern-discrimination in the vertebrates—III. Effective differences in width of visible striae for the monkey and the chick. *J. anim. Behav.* **6,** 169.

Klüver, H. (1933). "Behavior Mechanisms in Monkeys." Univer. Chicago Press, Chicago, Illinois.

Klüver, H. (1935). Use of vacuum tube amplification in establishing differential motor reactions. *J. Psychol.* **1,** 45.

Klüver, H. (1937). Certain effects of lesions of the occipital lobes in macaques. *J. Psychol.* **4,** 383.

Kohts, Nadie (1923). "Untersuchungen über die Erkenntnisfähigkeiten des Schimpansen." (In Russian.) Museum Darwinianum, Moscow. (German translation of summary, pp. 454–492.)

Leary, R. W. (1958). Homogeneous and heterogeneous reward of monkeys. *J. comp. physiol. Psychol.* **51,** 706.

LeVere, T. E. (1962). The effects of stimulus, response, and reward spatial contiguity on primate learning and the development of automated testing apparatus. Unpublished master's thesis, The Ohio State University.

McClearn, G. E., & Harlow, H. F. (1954). The effect of spatial contiguity on discrimination learning by rhesus monkeys. *J. comp. physiol. Psychol.* **45,** 391.

McConnell, D., Polidora, V. J., Friedman, M. P., & Meyer, D. R. (1959). Automatic reading and recording of digital data in the analysis of primate behavior. *IRE (Inst. Radio Engrs.) Trans. med. Electron.* **6,** 121.

Meyer, D. R. (1951a). Food deprivation and discrimination reversal learning in the monkey. *J. comp. physiol. Psychol.* **44,** 10.

Meyer, D. R. (1951b). The effects of differential rewards on discrimination reversal learning by monkeys. *J. exp. Psychol.* **41,** 268.

Meyer, D. R., & Harlow, H. F. (1949). The development of transfer of response to patterning by monkeys. *J. comp. physiol. Psychol.* **42,** 454.

Meyer, D. R., & Harlow, H. F. (1952). Effects of multiple variables on delayed response performance by monkeys. *J. genet. Psychol.* **81,** 53.

Meyer, D. R., Polidora, V. J., & McConnell, D. G. (1961). Effects of spatial S-R contiguity and response delay upon discriminative performances by monkeys. *J. comp. physiol. Psychol.* **54,** 175.

Miles, R. C. (1959). Discrimination in the squirrel monkey as a function of deprivation and problem difficulty. *J. exp. Psychol.* **57,** 15.

Murphy, J. V., & Miller, R. E. (1955). The effect of spatial contiguity of cue and reward in the object-quality learning of rhesus monkeys. *J. comp. physiol. Psychol.* **48,** 221.

Murphy, J. V., & Miller, R. E. (1958). Effect of the spatial relationship between cue, reward, and response in simple discrimination learning. *J. exp. Psychol.* **56,** 26.

Otteson, Mary I., Sheridan, C. L., & Meyer, D. R. (1962). Effects of stimulus-response isolation on primate pattern discrimination learning. *J. comp. physiol. Psychol.* **55,** 935.

Peterson, Marjorie E., & Rumbaugh, D. M. (1963). Role of object-contact cues in learning-set formation in squirrel monkeys. *Percept. mot. Skills* **16,** 3.

Polidora, V. J., & Fletcher, H. J. (1964). An analysis of the importance of S-R spatial contiguity for proficient primate discrimination performance. *J. comp. physiol. Psychol.* **57**, 224.

Polidora, V. J., & Main, W. T. (1963). Punched card programming and recording techniques employed in the automation of the WGTA. *J. exp. Anal. Behav.* **6**, 599.

Pribram, K. H. (1963). Reinforcement revisited: A structural view. *In* "Nebraska Symposium on Motivation" (M. R. Jones, ed.), pp. 113–160. Univer. Nebraska Press, Lincoln.

Pribram, K. H., Gardner, K. W., Pressman, G. L., & Bagshaw, Muriel (1962). An automated discrimination apparatus for discrete trial analysis (DADTA). *Psychol. Rep.* **11**, 247.

Riopelle, A. J. (1954). Facilities of the Emory University Primate Behavior Laboratory. *J. Psychol.* **38**, 331.

Riopelle, A. J., Wunderlich, R. A., & Francisco, E. W. (1958). Discrimination of concentric-ring patterns by monkeys. *J. comp. physiol. Psychol.* **51**, 622.

Schrier, A. M. (1958). Comparison of two methods of investigating the effect of amount of reward on performance. *J. comp. physiol. Psychol.* **51**, 725.

Schrier, A. M. (1961). A modified version of the Wisconsin General Test Apparatus. *J. Psychol.* **52**, 193.

Schrier, A. M., & Harlow, H. F. (1956). Effect of amount of incentive on discrimination learning by monkeys. *J. comp. physiol. Psychol.* **49**, 117.

Schrier, A. M., & Harlow, H. F. (1957). Direct manipulation of the relevant cue and difficulty of discrimination. *J. comp. physiol. Psychol.* **50**, 576.

Schrier, A. M., Stollnitz, F., & Green, K. F. (1963). Titration of spatial S-R separation in discrimination by monkeys (*Macaca mulatta*). *J. comp. physiol. Psychol.* **56**, 848.

Schuck, J. R. (1960). Pattern discrimination and visual sampling by the monkey. *J. comp. physiol. Psychol.* **53**, 251.

Schuck, J. R., Polidora, V. J., McConnell, D. G., & Meyer, D. R. (1961). Response location as a factor in primate pattern discrimination. *J. comp. physiol. Psychol.* **54**, 543.

Sheridan, C. L., Horel, J. A., & Meyer, D. R. (1962). Effects of response-induced stimulus change on primate discrimination learning. *J. comp. physiol. Psychol.* **55**, 511.

Spence, K. W. (1937). Analysis of the formation of visual discrimination habits in chimpanzee. *J. comp. Psychol.* **23**, 77.

Stollnitz, F. (in press). Spatial variables, observing responses, and discrimination learning sets. *Psychol. Rev.*

Stollnitz, F., & Schrier, A. M. (1962). Discrimination learning by monkeys with spatial separation of cue and response. *J. comp. physiol. Psychol.* **55**, 876.

Treichler, F. R., Horel, J. A., & Meyer, D. R. (1962). An assessment of automatically dispensed primate reinforcers. *J. exp. Anal. Behav.* **5**, 537.

Warren, J. M. (1953a). The influence of area and arrangement on visual pattern discrimination by monkeys. *J. comp. physiol. Psychol.* **46**, 231.

Warren, J. M. (1953b). Effect of geometrical regularity on visual form discrimination by monkeys. *J. comp. physiol. Psychol.* **46**, 237.

Warren, J. M. (1953c). Additivity of cues in visual pattern discriminations by monkeys. *J. comp. physiol. Psychol.* **46**, 484.

Warren, J. M. (1954). Perceptual dominance in discrimination learning by monkeys. *J. comp. physiol. Psychol.* **47**, 290.

Warren, J. M. (1956). Some stimulus variables affecting the discrimination of objects by monkeys. *J. genet. Psychol.* **88,** 77.

Weinstein, B. (1941). Matching-from-sample by rhesus monkeys and by children. *J. comp. Psychol.* **31,** 195.

Weinstein, B. (1945). The evolution of intelligent behavior in rhesus monkeys. *Genet. Psychol. Monogr.* **31,** 3.

Wunderlich, R. A., & Dorff, J. E. (1965). Contiguity relationships of stimulus, response, and reward as determinants of discrimination difficulty. *J. comp. physiol. Psychol.* **59,** 147.

Yerkes, R. M. (1943). "Chimpanzees: A Laboratory Colony." Yale Univer. Press, New Haven, Connecticut.

Chapter 2

Discrimination-Learning Sets[1]

Raymond C. Miles

Department of Psychology, The Ohio State University, Columbus, Ohio

I. INTRODUCTION

Experimental psychologists who began the controlled study of animal learning understandably were concerned with neutralizing all factors that might affect the function being studied. It was therefore customary to obtain "naive" subjects (animals with no laboratory learning experience) whose behavior presumably was not affected by previous training. Each experiment required its own allotment of subjects, since the same animals were never used more than once.

Experiments that formally investigated effects of previous learning

[1] This chapter was supported by Grant M-4890 from the National Institute of Mental Health, U. S. Public Health Service. The author wishes to acknowledge, with appreciation, the assistance of R. G. W. Miles.

generally were designed so that the reactions learned in one task would have a specifiable "transfer of training" effect on the second task. In general, previous training facilitated learning of a second task ("positive transfer") when the stimulus-response (especially response) patterns of the two tasks were essentially the same. When the response components were antagonistic, subjects were handicapped in learning the second task ("negative transfer"). With this simple and workable model for predicting transfer effects, most investigators were concerned only with transfer phenomena that could be schematized into discrete stimulus-response components.

There were data, however, which suggested that learning proficiency increases modestly but discernibly when subjects learn a number of similar tasks, even though each task comprises independent stimulus-response elements. Nonsense-syllable memorization was somewhat improved through practice on successive lists, even though each set of syllables differed from preceding lists (Melton & Von Lackum, 1941; Ward, 1937). Similarly, albino rats trained on a series of different maze patterns showed a slight increase in rate of learning as they progressed from problem to problem (Jackson, 1932; Marx, 1944; Wiltbank, 1919). Most of these studies, however, were not primarily concerned with general transfer effects; they used a series of similar training problems to obtain stable behavior for investigating some other issue.

II. BASIC LEARNING-SET PROCEDURES

Practically any type of learning procedure could be presented many times, with stimulus-response components varied for each individual problem, as described above. But the bulk of the research on learning-set formation has been based on the three procedures described in this section. Nonhuman primates have also formed learning sets for some of the problems described in Chapter 5 by French.

A. Discrimination Learning

More than 20 years ago, Harlow (1944) observed very rapid learning when he trained monkeys on 20 discrimination problems and 30 discrimination-reversal problems. He concluded "that once appropriate reaction sets have been formed in monkeys, these sets may be *transferred* from one pair of discrimination objects to another, making it possible for the subjects to meet a strict criterion for formation of a discrimination with a minimum amount of specific training" (Harlow, 1944, p. 11).

Nearly all primate learning-set research has used apparatus similar to the Wisconsin General Test Apparatus (WGTA; see Chapter 1 by Meyer *et al.*) that Harlow used, and most of it has used the learning task that he selected, object-quality discrimination. Stimulus objects can vary in several attributes (e.g., texture, height, brightness, form), making it possible to present many different discrimination problems. Each problem consists of a new pair of stimulus objects presented for a limited number of trials (usually six). One of the objects of each pair is arbitrarily designated "positive" and always covers food, usually a raisin or peanut. Thus, each subject is given many independent problems belonging to the same (object-quality discrimination) general class.

The first extensive investigation specifically designed to study general problem-to-problem transfer effects was done by Harlow (1949). The data established that training on many independent problems greatly enhances a monkey's proficiency at solving new problems of the same type. In this classic learning-set study, seven rhesus monkeys (*Macaca mulatta*) and a mangabey (*Cercocebus*) were trained in the WGTA on 344 independent discrimination problems. The first 32 problems, each 50 trials long, were divided into blocks of 8 problems; the next 200 discriminations, each 6 trials long, were divided into blocks of 100 problems; and the final 112, each 7, 9, or 11 trials long, were divided into blocks of 56 problems. Figure 1 summarizes the performance on trials 1 through 6 during successive blocks of problems. On the initial problems, per-

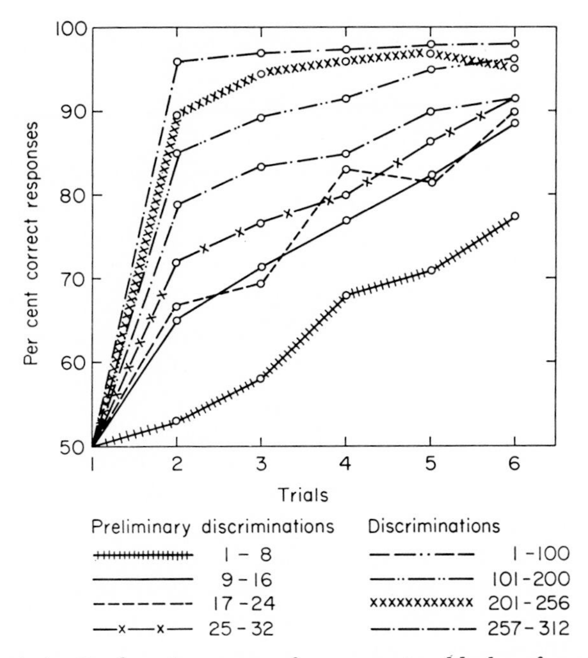

FIG. 1. Discrimination-learning curves for successive blocks of problems. (After Harlow, 1949.)

formance improved gradually over the first 6 trials, as it does on single discrimination problems. More important is the progressive increase in rate of learning successive individual discriminations until, at the end of the experiment, one trial was sufficient for almost perfect performance (95% correct on trial 2). Thus, these data exhibit two types of learning function, one representing changes in performance due to successive trials within problems (intraproblem learning), and the other representing changes in performance as a function of number of problems that have been presented (interproblem learning). Since the stimulus objects changed from problem to problem, recognition of specific objects or specific features of any object could not systematically facilitate solution of subsequent problems. Therefore, Harlow referred to this interproblem learning as the formation of a "learning set."

The data shown in Fig. 1 are, of course, group data. From such data there can be no assurance that learning by individual monkeys followed the same gradual course as did the group mean, and this is as true for interproblem learning as it is for intraproblem learning of the early problems. Conceivably, different monkeys could "catch on" at different times, each individual forming its learning set suddenly and thereby jumping from trial-and-error to insightful solution of new problems. But Schrier reports in personal communication that individual rhesus monkeys in a learning-set experiment (Schrier, 1958) progressed from slow to rapid solution of new problems in the gradual way that is indicated by the group curves.

B. Discrimination Reversal

A similar but even more orderly transition from initial, continuously improving intraproblem performance to the terminal discontinuous intraproblem learning function has been demonstrated by Meyer (1951a), using a discrimination-reversal-learning-set procedure (Harlow, 1944, 1949, 1950b). The nine rhesus monkeys in Meyer's experiment were first pretrained in the WGTA with 168 conventional 6-trial discrimination-learning problems. Then, in each of 108 reversal-learning problems, a pair of stimulus objects was presented for 6, 8, or 10 trials, followed by 8 more trials with responses to the previously positive object no longer rewarded, and choice of the previously negative object now rewarded. The number of prereversal trials varied unsystematically from problem to problem, so that the subject could not anticipate when reversal would occur. The initial discrimination-learning-set training made it possible to plot reduction in errors following the first reversal trial as the index of reversal-learning-set proficiency.

The reversal-learning-set data were summarized by Meyer in a way that clearly illustrates the intimate relation between initial performance and the entire course of interproblem improvement. An estimate of initial reversal-learning performance was obtained by plotting the mean number of errors as a function of reversal trials 1 through 8 on the first test day (four problems). All 108 problems were then divided into three blocks of 36 problems; mean data for each of these blocks depicted reduction in errors on trials 2 through 8 (following the first, "informative" reversal trial). These three curves were then plotted on the same graph with the first-day intraproblem learning function, but with the curve for each block displaced along the abscissa so that it was superimposed on the first-day function (Fig. 2). These transformed data exhibited a regu-

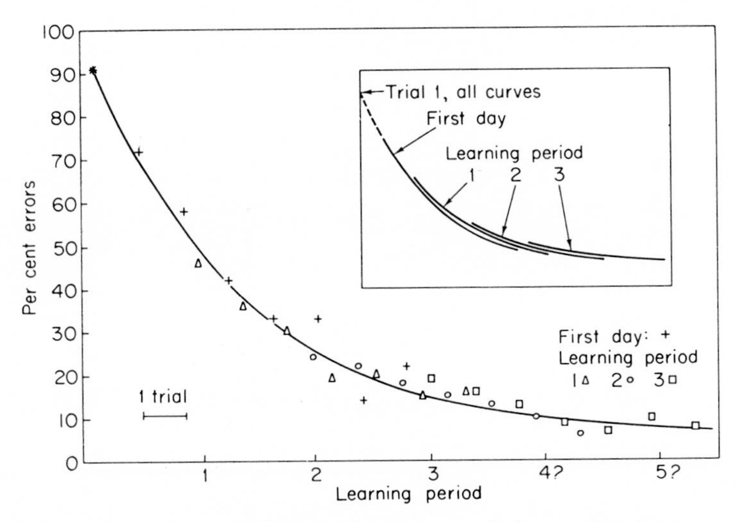

FIG. 2. A graph of the general reversal equation and the overlapping intraproblem functions from which it was derived. Each learning period includes 36 problems. (After Meyer, 1951a.)

lar and systematic function which could be represented by one general equation. Furthermore, equal amounts of training shifted the curve segments equal distances along the trial-unit abscissa. Such a degree of regularity is somewhat remarkable when one considers that each problem consisted of a different pair of stimulus objects and that the terminal intraproblem learning curve was quite different from the function exhibited at the beginning of reversal training.

C. Concurrent Discrimination

Another discrimination procedure, with some similarity to serial verbal- and maze-learning techniques, requires concurrent learning of several different discriminations. Rather than the usual method of presenting one pair of objects for a prescribed number of consecutive trials, a "list" of paired objects is presented for a prescribed number of "runs" with each pair appearing for just one trial during each run. The first stimulus pair is presented, then the second pair, then the third, and so on through the list, which is then repeated. A number of experiments by Leary (1957, 1958a, 1958b, 1958c, 1962) constitute the most thorough application of this "concurrent" technique. In one study (Leary, 1958a), nine rhesus monkeys, which had previously experienced the usual discrimination procedure, were given 10 runs through each list of nine stimulus pairs. A new list was used each day for 25 days. The results, grouped into 5-day blocks, are presented in Fig. 3. Correct object-selections in-

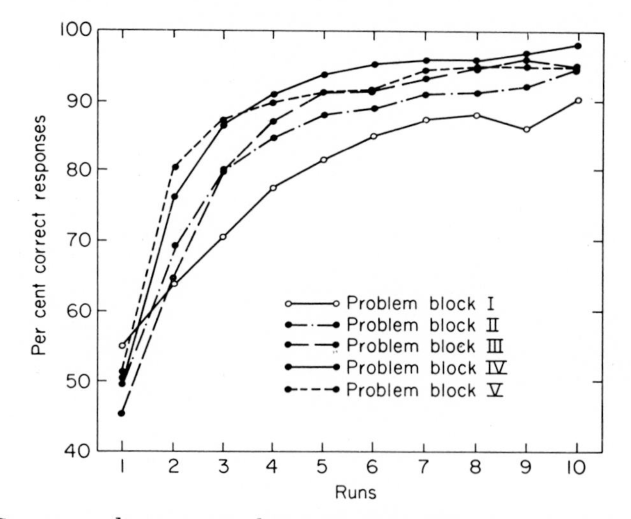

FIG. 3. Concurrent-discrimination-learning performance as a function of runs through lists of nine object pairs. Each problem block includes five lists. (After Leary, 1958a.)

creased regularly as a function of runs (intraproblem learning). The increase in correct object-selections as a function of test experience (blocks I through V) demonstrates the orderly interproblem development of a learning set.

Although the terminal level of proficiency in this experiment was high, other studies have shown that performance is better, especially on the early trials or runs, when discrete problems rather than concurrent-discrimination techniques are used (Darby & Riopelle, 1955; Hayes *et al.*, 1953b). When Hayes *et al.* (1953b) tested chimpanzees (*Pan*) with lists of 1, 5, 10, and 20 stimulus pairs, each subject's percentage of error on the second run increased with list length. The disproportionate difficulty of long lists presumably results from serial interference.

Only a few investigators have used concurrent techniques, but it appears that most of the parameters investigated with the conventional discrete-problem procedure could be explored using modified concurrent techniques. The concurrent-discrimination procedure could also be used to investigate with nonhuman primates such classic serial learning phenomena as retroactive and proactive interference.

III. PROCEDURAL VARIABLES

Of the variables to be discussed in this section, only one—problem length—is unique to learning-set procedures. The rest are included because they have been studied using learning-set techniques or because of their possible effects in learning-set experiments. Many other variables, equally relevant, are excluded only because they are treated extensively in other chapters. For example, the ability to form a learning set is directly related to both age (see Chapter 11 by Zimmermann and Torrey in Volume II) and phylogenetic standing (see Chapter 7 by Warren), and learning-set performance is strongly influenced by the important variables discussed in Chapter 1 by Meyer *et al.*

A. Problem Length

In most learning-set experiments, each discrimination problem is presented for 6 trials, a number which has proven to be adequate for reliable intraproblem learning while small enough for convenient presentation of many problems. An obvious practical concern, which also has theoretical implications, is the optimal number of trials to be used in each discrimination problem. Is learning-set formation most efficient when the total discrimination training consists of fewer problems with more trials per problem, or of numerous problems each with a relatively small number of trials?

As mentioned in Section II, A, Harlow (1949) used 50-trial problems during the early stages of his classic learning-set investigation, then switched to 6-trial problems. Discrimination performance improved over successive blocks of problems during both the early and the later stages of the study (Fig. 1). Braun *et al.* (1952) reduced the length of each problem to 3 trials after presenting eight 50-trial, twenty-four 10-trial, and thirty-five 6-trial problems, and found that their rhesus subjects took fewer total trials to equal the proficiency achieved by Harlow's subjects. Chimpanzees that had been trained to a criterion on 15 to 42 problems performed well on 2-trial problems (Hayes *et al.*, 1953a). But no improvement in performance was in evidence within 300 problems when a chimpanzee received 2-trial problems throughout the test series (Hayes *et al.*, 1953a), and rhesus monkeys given only 3-trial problems performed poorly in the early stages of training (Miller *et al.*, 1955).

Harlow (1959) interpreted these findings as indicating that there must be an appreciable amount of intraproblem learning if a learning set is to be formed. An efficient problem length should therefore include those trials on which maximum learning occurs. As a learning set is formed, intraproblem learning curves change from positively-accelerated, ∫-shaped, or linear functions, to negatively-accelerated functions. Harlow reasoned that the most efficient way to reach a given criterion of learning-set performance would be to start with long problems and to gradually reduce their length, so that trials on which less is learned would be excluded. During the final stages of learning-set training, when almost all intraproblem learning occurs between trials 1 and 2, two trials per problem would be most effective.

A test of these general predictions was the specific concern of an experiment (Levine *et al.*, 1959) in which number of trials per problem was the major independent variable. Thirteen rhesus monkeys, never before trained on discrimination problems, received 2,304 discrimination-learning trials; six subjects were given these trials in 3-trial problems, while seven subjects received 12 trials per problem. The percentage of correct responses on trial 2 was tabulated for the 12-trial animals. For the 3-trial subjects, data were taken from trial 2 of every fourth problem, i.e., from those trials that corresponded to trial 2 of the 12-trial problems. This made it possible to plot performances of both groups along an equivalent-unit abscissa (Fig. 4). Despite consistent training on problems of rather different lengths, the groups performed remarkably similarly throughout training.

This somewhat unexpected finding prompted Levine *et al.* (1959) to analyze data from other experiments using different problem lengths, which were comparable in that they had all used rhesus subjects 2 years old or older and relatively naive in discrimination learning; subjects had all been tested with a conventional discrimination-learning-set procedure, and each experiment had consistently used a single problem length throughout learning-set training. Four experiments met these specifications: an unpublished experiment by Meyer, Harlow, and Schiltz—6-trial problems, $N = 8$; a dissertation study by Levinson (1958)—4-trial problems, $N = 8$; an experiment by Schrier (1958)—4-trial problems, $N = 19$; and the first 32 problems of Harlow's (1949) original learning-set study—50-trial problems, $N = 8$. Comparisons among these data indicated that the effect of the same number of trials was less under the 50-trials-per-problem condition; learning-set development was similar with problems of all other lengths (ranging from 3 to 12 trials per problem) and discrimination proficiency was a function of total number of trials.

An investigation by Rumbaugh and McQueeney (1963), with the slower-learning squirrel monkey (*Saimiri sciureus*), indicated that learning each discrimination to a criterion (a technique resulting in a varying number of trials per problem) produces efficient learning-set performance. By having subjects learn each discrimination to a criterion of 20 correct selections within a 25-trial sequence, a level of proficiency equivalent to performance on 6-trial problems (Miles, 1957) was achieved in about one-third as many trials. On the other hand, a more recent study (Rumbaugh & Prim, 1963) reported no advantage for criterion training when procedures were directly compared. One group learned each problem to criterion, while two other groups were given the same number of trials (900) divided into either 6- or 60-trial problems. Similar learning-set performance was reported for all conditions.

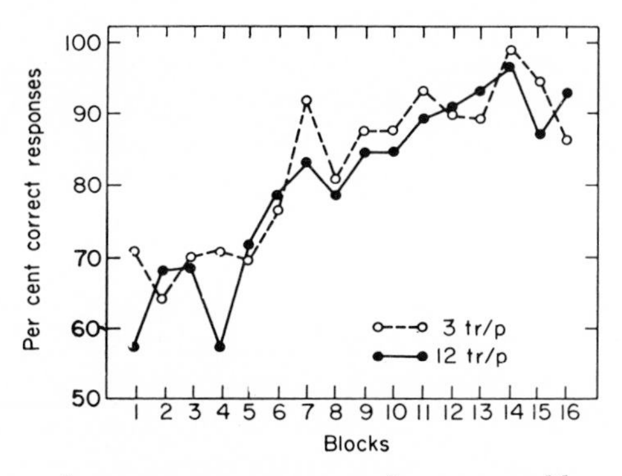

Fig. 4. Percentage of correct responses on the comparable second trials within each block of 144 trials. (After Levine *et al.*, 1959.)

An experiment by Sterritt *et al.* (1963) also revealed no difference in learning-set performance between six rhesus monkeys that were trained for 900 trials on 6-trial problems and six that received 900 trials of criterion training (20 correct in 25 selections). Level of performance depended only on the number of trials that had been given. As a final check, all subjects were then tested on 50 new 6-trial problems; the two groups showed equally proficient learning-set performance.

Although more representatives of other species should be tested before a general conclusion is drawn, it appears that proficiency of learning-set performance of rhesus monkeys depends primarily on total number of discrimination trials. The mass of rhesus data summarized by Levine *et al.* clearly justifies their statement that "$LS = f$ (trials) serves as a good approximation for the interval from 3 to 12 tr/p" (Levine *et al.*, 1959, p. 398).

B. Deprivation and Reward

Information about the possible influence of motivational variables, in addition to its theoretical value, should be considered when making even a cursory comparison of the various learning-set data. Although practically all learning-set research has used hungry subjects working for food reward, there have been minor but perhaps important variations in both deprivation and reward conditions.

1. Deprivation

Discrimination performance has been tested under lengths of food-deprivation ranging from 1 to 47 hours in rhesus monkeys (Meyer, 1951a) and from 1 to 20 hours in squirrel monkeys (Miles, 1959). The data are consistent in showing no differences in performance attributable to length of deprivation (see Chapter 1 by Meyer *et al.*). This makes it seem most unlikely that minor differences in deprivation had important effects on various learning-set functions. In addition, these findings decidedly restrict the generality of the thesis that variations in drive affect learned performance.

2. Reward

a. Quantity. Discrimination-learning-set performance does differ slightly but reliably as a function of changes in quantity of reward. Meyer (1951c) tested eight experienced rhesus monkeys with a reversal-learning-set procedure modified so that different amounts of food (one or three peanuts) could be given before and after the point of reversal. In half of the problems the number of peanuts found in the "correct" foodwell changed when the reward value of the two stimulus objects was reversed; in the other problems the amount of food did not change. Thus, four different reward conditions were studied: large reward before reversal, small after; small reward before, large after; large before, large after; and small before, small after. Errors were reliably reduced when the reward changed from small to large and were increased when a downward shift (from large to small) occurred. This differential effect on performance did not appear immediately but developed during the series of 64 problems, as if the monkeys were learning to discriminate quantity of food.

Three reward-quantity conditions were used in an experiment by Schrier and Harlow (1956). Eight cynomolgus monkeys (*Macaca irus*) were trained on 432 difficult 10-trial color-discrimination problems (see Chapter 1 by Meyer *et al.*). The reward conditions (1, 2, and 4 food pellets) varied from problem to problem so that each subject had equal experience with all three conditions. Discrimination performance varied directly with amount of food.

Schrier's (1958) comprehensive learning-set investigation on this topic is noteworthy because of the relatively large number of subjects (24 naive rhesus monkeys) and because two methods of comparison were used over a wide range of reward-quantity conditions (1, 2, 4, and 8 food pellets). Subjects were divided into two groups: a "shift" group in which each subject experienced all reward-quantity conditions, and a "nonshift" group in which four different subgroups each experienced only one reward condition throughout the entire series of 160 four-trial object-discrimination problems. A positive linear relation was found between mean performance of the "shift" group and amount of food. On the other hand, a nonlinear relation was found for the "nonshift" group; there was little difference in discrimination performance of these subjects for the 1-, 2-, and 4-pellet amounts, but subjects receiving 8 pellets performed appreciably better. All learning-set research, other than that directly concerned with effects of amount of reward, has used a single reward condition throughout each experiment. On the basis of Schrier's results, it appears safe to assume that variations in amount of reward have not influenced the various learning-set functions as much as might have been expected.

b. Quality. There has not been any learning-set research on possible effects of differences in quality of reward. It should be noted, however, that the choice of reward is often determined by the subject's behavior. A given subject sometimes refuses a particular food, forcing the experimenter to use another kind of reward (see Chapter 1 by Meyer *et al.*).

C. Stimulus Generalization

The effect on discrimination efficiency of the magnitude of difference in a single cue attribute was studied by Grandine and Harlow (1948). Their subjects, 11 rhesus monkeys and a mangabey, had previously formed discrimination- and reversal-learning sets. Grandine and Harlow measured generalization by a two-choice discrimination technique using sets of stimulus objects which varied either in height or in brightness and saturation. Four sets of objects (triangular, rectangular, inverted-T-shaped, and L-shaped) each varied in *height* to represent five points along a continuum from 1 inch (Stimulus 1) to 3 inches (Stimulus 5); four other sets of objects each varied in *brightness* and *saturation* to represent five points along a continuum from one part colored paint plus two parts white (Stimulus 1) to all colored paint (Stimulus 5). Each discrimination problem consisted of two stages: (1) A single stimulus object was presented for 25 consecutive training trials. In half of the problems, this singly-presented object was designated "positive," and pushing the object aside revealed food in an underlying foodwell; in the other problems, the training stimulus was designated "negative," and object displacement was not rewarded. When a subject failed to displace a negative object within 5 seconds, the experimenter ended the trial by displacing the object and exposing the empty foodwell. (2) After the 25 training trials on each problem, the subject received 25 generalization test trials in which the training object was paired with a new test object on the same stimulus continuum. Stimulus 5 (the tallest or darkest) was a member of every test pair, and the order of problems proceeded from the pair with the greatest difference between stimuli (5 versus 1) to the pair with the least difference (5 versus 4). Percentage of first-trial errors under each of the four difference-magnitude conditions was used as a measure of generalization; an error was defined as a response not in accord with the reinforcement contingences of training (i.e., selecting a previously negative object or not selecting a previously positive object). The functions in Fig. 5 clearly show that generalization was a function of the physical disparity between paired stimulus objects. Although both rewarded and nonrewarded training produced generalization gradients, it should be pointed out that these gradients were not identical: the gradient after nonreward was noticeably steeper.

Although it had been commonly assumed that the stimulus generalization phenomenon of classical conditioning played a part in discriminations involving more complex stimuli, there had been little research evidence in support of such an extrapolation. This excellent study clearly showed that discrimination of complex stimuli involves an analogous process. Although the exact shape of the generalization gradients cannot be deduced because difference magnitudes were not measured in j.n.d. units, decreased magnitudes of difference between paired stimulus objects clearly resulted in less-efficient discrimination performance. Even if one assumes that "generalization" represents only a lack of discrimination between two stimuli (Lashley & Wade, 1946), the fact remains that there was a systematic relation between discrimination "errors" and the physical disparity of paired test objects.

In an earlier discrimination-generalization study, Harlow and Poch (1945) presented 12 problems to 11 rhesus monkeys and a bonnet macaque (*Macaca radiata*). In each problem, the animals were trained to discriminate between members of a pair of stimulus objects (to a criterion of 45 correct in 50 consecutive trials), and then were given 12 generalization tests in which choice of either stimulus object was rewarded. The members of a training pair differed either in one attribute (color, form, or size) or in two attributes (color and form). One or both members of test pairs differed from training pairs in one of these three attributes. Generalization, measured as a reliable deviation from chance responding, occurred to some test pairs under each of these conditions.

But differential responding to test pairs was always less than to training pairs, whether or not the changed attribute was one in which the training stimuli differed. This nonequivalence of training and test pairs indicates that discrimination performance was controlled by more than one stimulus attribute. The monkeys generalized color discriminations more readily than form or size discriminations.

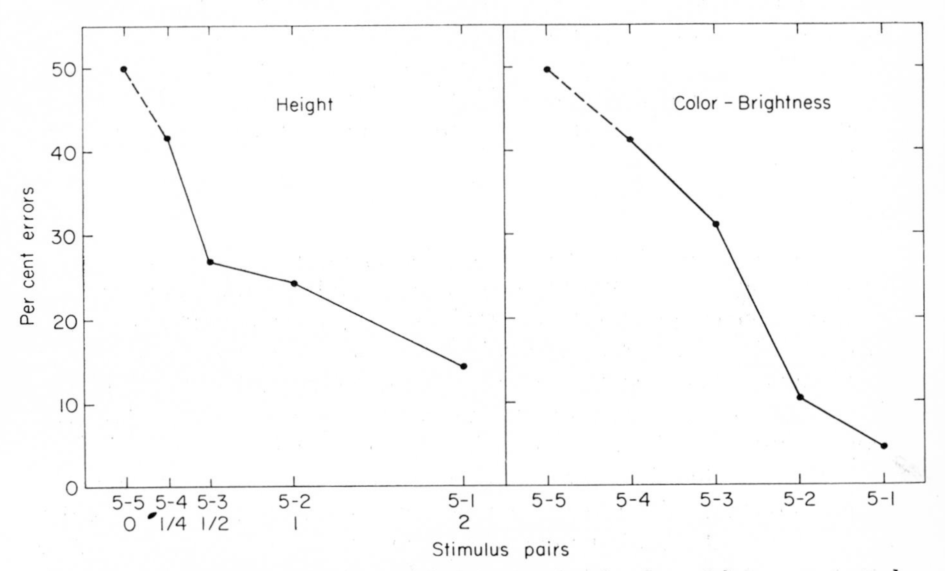

FIG. 5. Percentage of first-test-trial errors under the four difference-magnitude conditions for height and color-brightness stimuli. Chance performance on 5–5 is hypothetical. (After Grandine & Harlow, 1948.)

The results of Harlow and Poch's experiment agree with other studies of the discriminability of stimulus differences (see Chapter 1 by Meyer *et al.*) in suggesting that color is a dominant cue, but that monkeys use cue information from more than one, probably many stimulus attributes in the solution of object-quality-discrimination problems. Perhaps color is most effective because color has three dimensions (hue, saturation, and brightness), all of which may vary and may contribute to the discriminability of the difference between two stimulus objects.

IV. CHARACTERISTICS OF LEARNING-SET BEHAVIOR

A. Stability

As a learning set is formed, performance improves in an orderly and systematic way. It should therefore be possible to use data from one block of problems as a basis for predicting a subject's level of performance on the following block, or perhaps even for predicting the entire course of a subject's learning-set attainment. The apparent consistency of learning-set performance was the specific concern of a study by Miller *et al.* (1955), who tested 23 rhesus monkeys on 567 three-trial object discriminations. The data from trials 2 and 3 were grouped into six blocks of 96 problems for purposes of analysis. Split-half reliability coefficients between odd and even days were low for the first block of problems, but ranged from 0.80 to 0.96 on problem blocks 2 through 6. Correlations between numbers of errors in the five pairs of adjacent problem blocks (1 and 2, 2 and 3, etc.) were: for trial 2–0.16, 0.82, 0.79, 0.83, and 0.95; and for trial 3–0.56, 0.66, 0.91, 0.93, and 0.93. It appears that, after initial variability, learning-set performance is fairly stable.

B. Retention

Because each of hundreds of stimulus pairs is presented for only a brief sequence of consecutive trials, conventional discrimination-learning-set formation could represent merely a temporary facilitation of performance. The concurrent-discrimination results (Section II, C), however, suggest some permanent acquisition, as several minutes usually elapse between repeated presentations of a particular stimulus pair. Riopelle and Churukian (1958), who were interested in relatively short-term recall, tested four rhesus monkeys on a variety of problems including object-quality discriminations with intertrial intervals of either 10 or 60 seconds. During each test session, performance was about 5% better for the shorter intertrial interval, but this effect was reversed when subjects were tested on the same problems 24 hours later. It should be emphasized that all differences were very small, and that neither the intertrial interval nor the 24-hour time lapse had any appreciable effect on discrimination proficiency.

Retention over longer intervals has also been studied. Braun *et al.* (1952) found no loss in learning-set performance (i.e., in the speed of learning new discriminations) after about 8 weeks of no training. Specific stimulus-object discriminations are also retained surprisingly well. Strong (1959) trained four naive rhesus monkeys on 72 pairs of stimulus objects to a criterion of 32 correct in 36 responses. Retention tests given 30, 60, 90, and 210 days after criterion was met showed that a high level of accuracy (90% correct responses) was maintained over the 210-day period. Although use of a criterion assured a fairly high level of learning on each discrimination, this almost perfect recall (ranging from 86 to 92% correct) of 72 separate discriminations is most impressive. These results support other, less direct, retention data in implying that discrimination-learning-set performance represents a fairly long-term process rather than a temporary response set.

C. Systematic Responding

The learning curves used to illustrate development of a learning set usually represent mean discrimination performance within relatively large blocks of problems. This somewhat gross description is useful for showing general learning trends but provides minimal information about underlying behavioral processes. The wide variety of the events that influence discrimination-learning-set performance of rhesus monkeys suggests that stable learning-set functions based on summarized data may represent the combined effects of many different modes of responding.

1. ERROR-FACTOR ANALYSIS

By means of a rather complicated data analysis, Harlow (1950a) showed that errors made during learning-set training are not just "accidental" misses but in many instances represent systematic response tendencies that are inappropriate to solution of the problems. This general approach is reminiscent of some earlier analyses undertaken to better comprehend maze-running behavior of rats by isolating response patterns such as "forward-going tendencies," "goal anticipatory reactions," "centrifugal swing," etc. (Maier & Schneirla, 1935). Harlow (1950a) identified four systematic response tendencies, "error factors," which were related to selection of negative stimulus objects in the first 200 six-trial problems labeled "Discriminations" in Fig. 1. Harlow (1950b) analyzed the data from the 112 discrimination-reversal problems that followed.

a. Stimulus perseveration. Selection of a stimulus object, after response to that object was unrewarded on the first trial, was labeled "stimulus perseveration" error. These errors were attributed to preference for, or avoidance of, certain stimulus objects. Although it is not yet possible to specify conditions responsible for stimulus-perseveration error, some observational data and the results of object-preference research (Harlow, 1959; Menzel, 1962) suggest that monkeys tend to avoid large, unstable, or many-pointed, bright, metallic stimuli. Harlow (1959) also suggested that generalization from stimulus qualities of previously positive and negative objects would influence preferential responses to particular objects. Both the total number and the intraproblem persistence of stimulus-perseveration errors decreased during learning-set training (Fig. 6).

b. Differential cue. Before the trial on which the two stimulus objects first change position, the outcomes of object-displacement could be associated either with the right-left position or with the stimulus qualities of the chosen object—a situation referred to by Spence (1936) as "ambiguous reinforcement." To respond correctly on the first trial following positional change, the subject must respond on the basis of object quality irrespective of position. Selection of the negative object on this critical trial was referred to by Harlow as a "differential cue" error, and these errors were attributed to "ambiguous" positional reinforcement on previous trials. When the number of errors made on a particular differential-cue trial (e.g., trial 2) is divided by number of errors on comparable trials of sequences not involving positional change, the resulting comparable-trial error (CTE) ratio indicates the contribution of differential-cue errors. Early in training, two immediately-preceding ambiguous rewards resulted in a higher CTE ratio than did one, but the

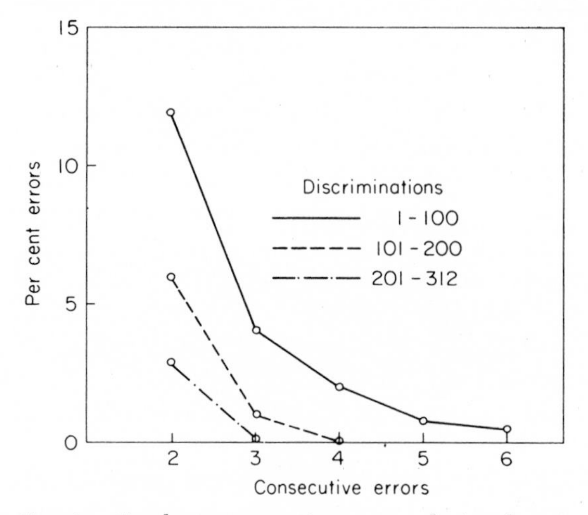

FIG. 6. Reduction in stimulus-perseveration errors during learning-set formation. (After Harlow, 1959.)

number of differential-cue errors decreased as a function of experience until, after 200 problems, the CTE ratio was very close to 1.0 (Fig. 7). These results suggest that object-quality-discrimination performance of experienced subjects was no longer impaired by irrelevant relations of reward to left-right position. Differential-cue errors reappeared in reversal learning (Harlow, 1950b).

c. Response shift. Selection of the negative member of a stimulus pair after consistent displacement of the positive object was called "response shift" error. It was attributed by Harlow to a tendency, perhaps innate, to explore or manipulate both stimulus objects in a discrimination problem. Original evidence for this persistent type of error was found when Harlow's (1950a, 1950b) data analyses revealed that more discrimination errors were made on trials 4, 5, and 6 when all previous selections were correct than when an initial incorrect and subsequent correct object-selections had occurred.[2] The finding that discrimination performance is less efficient after displacement of a singly-presented positive object than after displacement of a negative object (Section V, A, 5) gave further evidence for the occurrence of "response shift" errors.

An interesting phenomenon was revealed when data from Schrier and Harlow's (1956) experiment on quantity of reward (Section II, B, 2, *a*) were subjected to an error-factor analysis. Frequency of response-shift errors was inversely related to quantity of food, and such errors progressively decreased with practice under all reward conditions. But an analysis of errors on trial 2 following correct responses on trial 1 showed

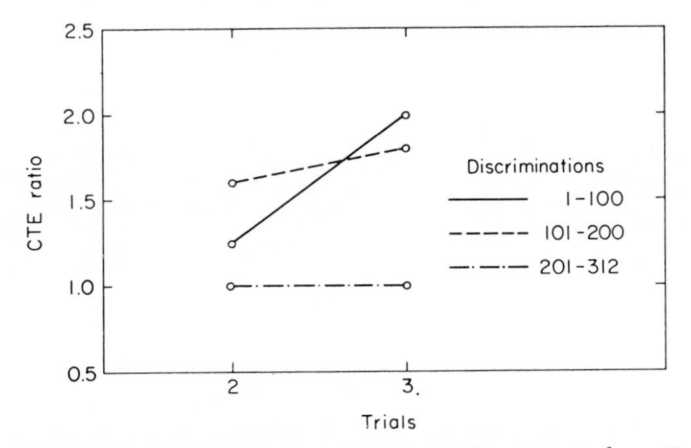

FIG. 7. Reduction in differential-cue errors during learning-set formation as indicated by decrease in CTE ratio. (After Harlow, 1950a, 1959.)

an interesting relation between amount of reward and the temporal course of response-shift-error frequency. While number of response-shift errors under four-pellet and two-pellet conditions remained fairly constant throughout the experiment, errors on trial 2 under the one-pellet condition progressively increased as a function of learning-set experience (Fig. 8). In other words, it appeared that subjects progressively learned to make response-shift errors on trial 2 under the low-reward condition. The analysis showed no systematic effects of amount of food on either differential-cue or position-habit errors, and these results were taken as further evidence that the several error-producing factors function independently.

d. Position habit. Repetitive responses to either the right or the left foodwell, regardless of position of the positive object, were attributed to the particular positional preferences of individual monkeys. "Position

[2] Moss and Harlow (1947) indicated that this analysis had already been done when they wrote their paper.

habit" error was defined as a significant excess of incorrect responses to either the right or left position. Position preference generally had a negligible effect upon discrimination-learning efficiency (Harlow, 1950a).

2. Trial-to-Trial Response Patterns

A systematic tabulation of trial-to-trial behavior has revealed response patterns that are positively related to correct object-selections rather than to errors. There are four different patterns of behavior that can repeatedly occur from trial n to trial $n + 1$ within a two-choice discrimina-

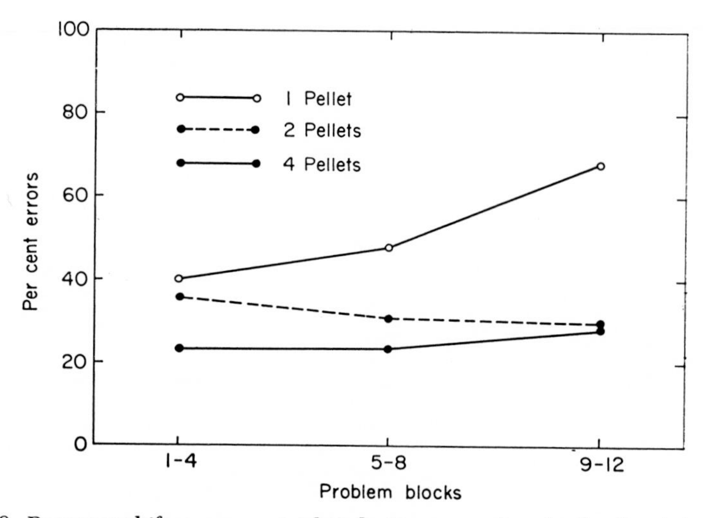

Fig. 8. Response-shift errors on trial 2 during successive thirds of training for the three amounts of reward. (After Schrier & Harlow, 1956.)

tion problem. Two of these are defined in terms of spatial location (on trial $n + 1$, the subject may uncover either the same foodwell as on trial n or the other foodwell), and two are defined in terms of stimulus objects (the subject may push aside either the same object or the other object). Each of these four trial-to-trial response patterns can result in either reward or nonreward.

To investigate trial-to-trial behavior, I have analyzed the performance of seven squirrel monkeys on 1,000 six-trial discrimination problems. Figure 9 shows the median frequencies of each pattern of behavior in 250-problem blocks. Frequencies of shifting position and of pushing aside the same object increased reliably with practice.

In the case of a hypothetical subject incapable of learning, the various patterns of trial-to-trial behavior would be correlated with "chance" performance (50% correct choices) on trial $n + 1$. Figure 10 shows the actual relation between each of the four response patterns and percentage of correct choices on trial $n + 1$. Each behavior pattern was positively related to correct object-selections, and the trends increased as training progressed. The percentage of correct responses was lowest when the

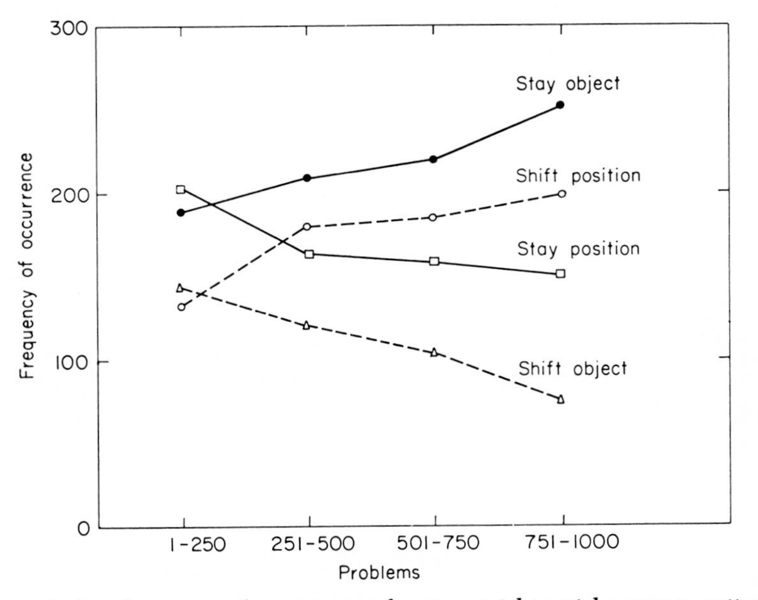

FIG. 9. Median frequency of occurrence of various trial-to-trial response patterns.

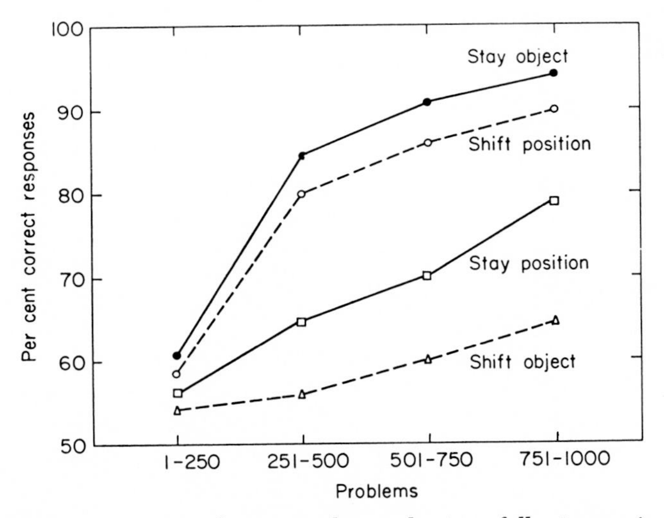

FIG. 10. Mean percentage of correct object-selections following various trial-to-trial response patterns.

subject changed an object-selection, a result which is probably related to the tendency to explore both objects.

A similar analysis was then carried out to determine the cumulative effect of consecutive (i.e., "perserverative") rewarded or nonrewarded selections of the same object upon discrimination performance on the following trial. Discrimination efficiency was directly related to the number of consecutive selections of either positive or negative objects (Fig. 11). These results indicate that the subjects benefited from cumu-

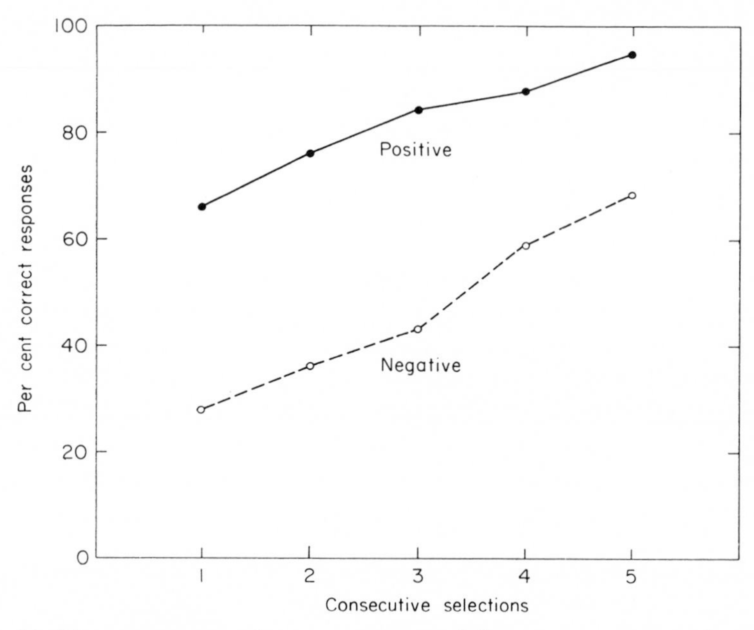

FIG. 11. Mean percentage of correct responses following consecutive selections of positive and negative stimulus objects.

lative "informative" effects of either rewarded or nonrewarded object-selection in proportion to how often these events occurred. The decidedly lower performance level following nonrewarded selections could be attributed to the fact that the subject, in order to make a correct choice, must select a nonpreferred object.

The results of these rather simple analyses indicate, at least for squirrel monkeys, that all of the trial-to-trial behavior patterns are relevant to procuring reward. The fact that these systematic response patterns all resulted in higher probability of reward as training progressed indicates that position perseveration, position alternation, object perseveration, and object alternation may all represent behavior that is shaped through the procurement of reward.

V. SPECIAL LEARNING-SET PROCEDURES

A number of investigators have modified the usual discrimination-learning-set procedure in various ways. Data from these experiments will undoubtedly contribute to our eventual understanding of learning-set formation by clarifying particular aspects of the learning process and indicating some of the complexities involved in learning how to learn.

As a discrimination is learned, the stimuli that are consistently related to reward and nonreward gain control of behavior. When a subject is tested on a long series of independent but similar problems, its behavior can also become controlled by events that are related to reward and nonreward in every problem in the series. Events such as the outcomes of responses and the introduction of new stimulus objects can become cues signifying the test condition currently in effect. It is therefore possible to specifically design learning-set procedures in order to study various single-problem and problem-to-problem learning processes.

A. Outcomes as Cues for Interproblem Learning

Throughout the lengthy course of learning-set training a subject repeatedly experiences two consistent outcomes—reward following displacement of a positive object, and nonreward following displacement of a negative object. Almost-perfect second-trial performance shows that either outcome enables a sophisticated subject to solve a discrimination problem. Reinforcement, or strengthening, of the immediately preceding response is the function usually attributed to reward, but the presence or absence of "expected" reward can also function as a discriminative cue. Nonreward is a highly informative event for a subject that has attained a high level of learning-set proficiency, as are other outcomes for subjects specially trained with them. Thus, the procedures reviewed in this section are like the basic learning-set procedures (Section II) in at least one respect: response on trial n produces an outcome (food, empty foodwell, or some other stimulus) that can serve as a cue for correct choice on trial $n + 1$.

An interesting learning function obtained by Warren (1954) shows that the cue function of reward develops during a series of independent reversal problems until object-selection is virtually independent of the strengthening effect usually attributed to reward. This particular study is noteworthy because of the wide variation in number of rewarded responses (ranging from 1 through 32) made before reversal, and because the reversal procedure was introduced at the very beginning of training rather than after the customary pretraining series of conventional discrimination-learning problems. Each rhesus macaque was tested on 108 object-quality-discrimination-reversal problems. Prereversal training continued on each problem until displacement of the positive object had been rewarded 1, 2, 4, 8, 16, or 32 times, whereupon reward values were reversed and food was obtainable under the previously negative object for 10 postreversal trials. An increase in number of prereversal rewards impaired postreversal performance on the first 36 problems; subjects made about 21% fewer errors on the second postreversal trial after 1 prereversal reward than after 32 rewards. On the final 36 problems, however, there were only 5% fewer postreversal errors after 1 than after 32 rewards, a difference which was not statistically significant.

Frequency of prereversal reward also was varied by Leary (1962), who tested nine experienced rhesus monkeys on concurrent discrimination learning (Section II, C). Nine lists of eight pairs of stimulus objects were presented to each subject; reward values of paired objects were reversed after either 1, 3, or 6 prereversal runs. The monkeys made reliably fewer errors after 1 than after 3 or 6 prereversal runs. This effect might have decreased with further training, since the total discrimination-reversal experience of these subjects was considerably less than that of Warren's (1954) monkeys.

Results of an investigation by Riopelle *et al.* (1954) indicate that the relative importance of the reinforcing and cue functions of reward changes during the formation of object-quality-discrimination-learning sets. Four groups of three naive rhesus monkeys were tested on 250 six-trial discrimination problems. The procedure differed from conventional learning-set training only in the methods used to signify the positive stimulus object on the first trial of each problem. Each of the four groups was tested under a different first-trial condition: For group I, both foodwells were baited on the first trial so that food was obtainable no matter which object was chosen, and continued selection of the initially chosen object was rewarded on trials 2 through 6. For group II, the choice on the first trial was defined as "correct," as in group I, but in this case displacing the object revealed a marble in the foodwell, and continued selection (trials 2–6) of the initially chosen object was rewarded with food. For group III, the choice on the first trial was always "incorrect" (no food in either foodwell), and selection of the alternative object was rewarded on trials 2 through 6. For group IV, the choice on the first trial revealed a marble, and the subject had to select the alternative object on subsequent trials in order to obtain food reward. All groups showed an increase in discrimination efficiency with increasing test experience, i.e., consistent experience with any of the four first-trial conditions was sufficient for learning-set development. As would be expected, subjects that had to change their initial object-choice (groups III and IV) were reliably less proficient than those (groups I and II) that had to continue to select an initially chosen (preferred) object. The reinforcing function of food reward on the first trial was most important in the early problems in the series; by the end of the experiment, a correct selection on trial 2 more often followed a *nonrewarded* than a rewarded response on trial 1. Thus, the informative function of first-trial outcomes became progressively more important as test experience increased. In this connection, it is interesting that a marble under the positive object on the first trial produced as efficient discrimination performance as did food.

Bowman (1963) studied the effects of informative response outcomes on object-quality-discrimination-learning-set performance under four reward-percentage conditions. Food was placed under the positive object on 3 trials of each problem; total number of trials per problem was either 3 (100% reward), 4 (75% reward), 6 (50% reward), or 12 (25% reward). Eight experienced rhesus macaques received 400 problems under each reward condition, 200 with discriminable stimuli (colored foodwell covers) located beneath the objects and 200 without such stimuli (identical gray covers over both foodwells). Interproblem learning occurred either with or without colored covers under 100% and 75% reward conditions, but at 50% and 25% reward there was no interproblem increase in discrimination efficiency without colored covers. With colored covers, performance under the two smaller reward percentages did improve with practice, indicating that the information provided by the colors was used. Analysis of intraproblem performance on the last 40 problems for each intermittent-reward condition gave some evidence that subjects had learned to use either a positive or a negative color on a nonrewarded first trial as a cue for correct object-selection on trial 2. Performance on subsequent trials was better with than without colored covers, indicating that informative response outcomes aided the subject in sustaining intermittently-rewarded "correct" responses.

1. OBSERVATIONAL CUES

Darby and Riopelle (1959) showed that the informative outcome does not have to result from the subject's own response. Experienced rhesus macaques can achieve a respectable degree of discrimination proficiency merely by observing the outcome of another monkey's object-selection. The first trial of each 6-trial problem was a demonstration trial and the other five were discrimination trials. On half of the problems, the object selected by the observed monkey was positive (both foodwells were baited on the demonstration trial so that pushing aside either object procured food), and the observing monkey found food on discrimination trials only if it selected the object that had been displaced on the demonstration trial. Presented alternately with these problems were problems introduced by a demonstration trial on which the object-selection made by the observed monkey was "incorrect" (neither foodwell was baited), and the observing monkey then obtained food only by selecting the object not displaced on the demonstration trial. The four subjects, each of which was given 500 problems, all learned on the basis of observational cues and by the end of the experiment were making close to 75% correct selections on the first discrimination trial of each problem.

2. Elimination of Irrelevant Response Tendencies

Almost all of the learning-set studies described so far have been primarily concerned with the development of responses to cues that are relevant to procuring reward. A correlative occurrence, usually presumed to be of equal importance, is the elimination of modes of behavior that have no relevance to problem solution (e.g., responding on the basis of positional reward during object-quality training, "attending" to irrelevant features of the test tray, etc.). Although it is extremely difficult to measure these behavioral changes, a number of investigators assume that these processes are basic to the development of a learning set. There is at least one theoretical account of learning-set formation (discussed in Section VI, A) that suggests that observed functions result from the elimination (inhibition) of behavior that produces errors.

An intriguing experiment by Meyer (1951b) showed that rhesus monkeys, extensively trained on conventional discrimination-reversal problems, learned to disregard the previously relevant cue of nonreward when this event became a misleading cue. By the end of the reversal training described in Section II, B, the reversal cue (a nonrewarded selection of a previously positive object) resulted in selection of the alternative object about 80% of the time (Fig. 2). About a year later, 200 eight-trial "reversal cue" problems were given to eight of the nine original subjects. A discrimination was established at the beginning of each problem, the reward values were reversed for a *single* "reversal cue" trial (either the fourth, fifth, or sixth trial in the eight-trial problem), and then the initially positive object was again positive for the remaining trials. As would be expected, discrimination performance was severely disrupted by the "reversal cue" trial, which had indicated during discrimination-reversal training that reward values would continue to be reversed on subsequent trials. Most important was the development of a "set" to ignore the "reversal cue," so that discrimination efficiency on subsequent trials became less and less affected by this now-irrelevant event (Fig. 12). Subjects then were given 20 reversal-relearning problems according to the original discrimination-reversal procedure. The reversal-relearning and the original reversal-learning functions were very similar (Fig. 13), suggesting that reversal-learning ability was not impaired by experience with an irrelevant reversal cue but that the irrelevant-reversal-cue experience had destroyed the original reversal-learning set. The near identity of the two functions indicates a complete loss of the learned repertoire essential for efficient discrimination-reversal performance.

Results of the "reversal cue" phase of Meyer's experiment clearly indicate that rhesus macaques progressively learn not to be affected by an irrelevant intraproblem event. A similar interproblem phenomenon was

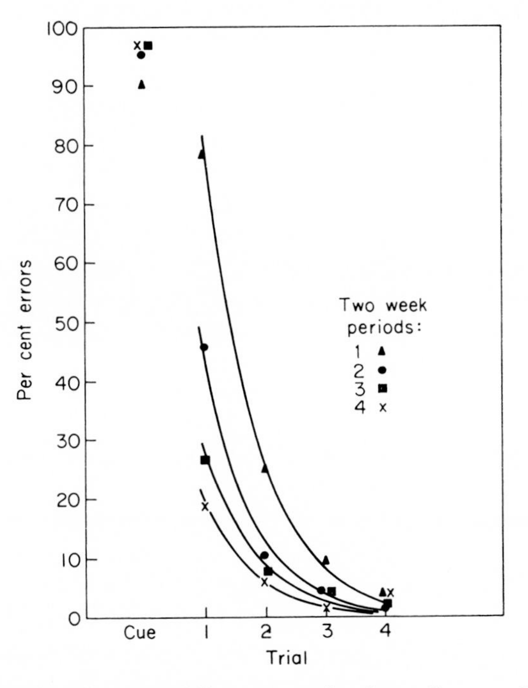

FIG. 12. Intraproblem learning following an "irrelevant" reversal-cue trial. Each 2-week period represents 50 problems. (After Meyer, 1951b.)

demonstrated by Riopelle (1953); subjects learned to ignore irrelevant problem-to-problem similarities, a process referred to by Riopelle as "transfer suppression." Particular stimulus-reward relations in one problem are irrelevant to solution of other independent problems, and learning to choose a *particular* stimulus object in one problem may actually interfere with solving later problems. To analyze the influence of previous stimulus-reward relations on discrimination performance throughout learning-set training, Riopelle used a procedure designed to maximize the amount of interfering interproblem "transfer." Four rhesus macaques received five 6-trial discrimination problems on each test day. After completion of the five problems, a sixth problem was presented using the stimulus pair from either the first or the fourth problem but with reward values reversed. Testing continued until each subject had received 315 new and 63 reversed problems. During early training, performance on reversed problems was considerably poorer than new-problem performance; this initial difference in performance progressively diminished during the experiment until the reversed-problem performance and the new-problem performance were practically identical. Degree of retardation in performance on reversed problems was taken as a measure of

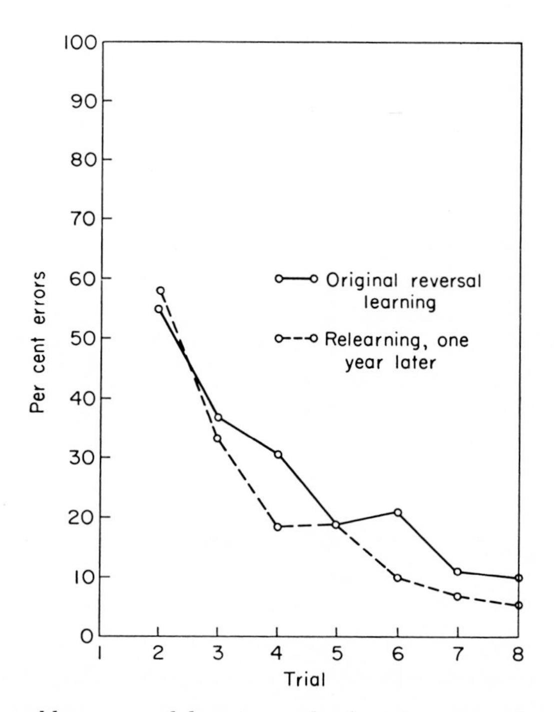

FIG. 13. Intraproblem reversal learning and relearning curves based on 20 problems each. (After Meyer, 1951b.)

transfer effects from previous experience with particular stimulus objects, and Riopelle interpreted the results as implying that a decrease (or "suppression") of irrelevant problem-to-problem transfer accompanies the formation of a discrimination-learning set. Thus, subjects apparently learn to become progressively less and less affected by irrelevant interproblem similarities of stimuli, until discrimination performance on independent problems becomes free from the influence of outcomes associated with particular stimulus objects in preceding problems.

3. Conditions that Facilitate Learning-Set Formation

Most learning-set procedures involve training on many independent problems. But at least two experiments have shown that such training is not necessary to develop the ability to solve new problems in one trial.

Schusterman (1962) used a repeated-reversal procedure (see Chapter 5 by French and Chapter 7 by Warren) to train chimpanzees to keep selecting the same object when the response on the previous trial was rewarded and to select the other object when the response on the previous trial was not rewarded. This, of course, is the optimal strategy for both discrimination and discrimination-reversal learning. It is described by Levine (1959) as "Win-stay-Lose-shift with respect to the object." (Various strategies or "hypotheses" that appear during learning-set formation are discussed further in Chapter 3 by Levine.) Schusterman trained four experimentally naive chimpanzees on repeated reversals of three discriminations—800 trials with the first pair of objects, 300 trials with the second, and 200 with the third. One member of a pair covered food until the subject made 12 consecutive correct responses, then reward was shifted to the other object until the criterion was attained, etc. After this rather extensive discrimination-reversal training, subjects were tested on 180 six-trial object-quality-discrimination problems. They immediately exhibited proficient learning-set performance. Subjects made about 88% correct object-selections on trial 2 during the first block of 30 problems, indicating that they were able to obtain on the first trial all the information necessary for problem solution.

In a similar study, Riopelle (1955) first trained five naive rhesus monkeys on 10 discrimination problems and then used only four stimulus objects to construct 216 problems, each of which was learned to a criterion of five consecutive correct responses. (Four objects can be paired six ways; 12 different problems were obtained by reversing the reward values of each pair, and the 12 problems were repeated 18 times for a total of 216 problems.) When the subjects then were tested with a discrimination-learning-set procedure, a high level of proficiency was exhibited on the first test day.

These experiments indicate that special training on a Win-stay-Lose-shift strategy, by means of procedures arranged so that particular stimulus properties are not consistently relevant, considerably facilitates the development of a discrimination-learning set. An earlier experiment by Riopelle (1953) shows that prolonged Win-stay-Lose-shift training is not sufficient when specific stimulus-reward relations are maintained throughout. Four rhesus monkeys received more than 2,000 training trials on six difficult discriminations before being tested on a series of independent 6-trial problems. Despite thorough adaptation and extensive discrimination training, the early learning-set performance of these animals was similar to that of four rhesus monkeys that had not received prior training (Miles & Meyer, 1956).

4. Restriction of Learning to Displaced Objects

A recent experiment (Lockhart *et al.*, 1963) shows that monkeys able to solve problems in one trial learn practically nothing in that trial about objects that they do not actually push aside. Eight pig-tailed macaques (*Macaca nemestrina*), with previous experience in learning-set experiments, were given 320 six-trial problems. In each problem, a pair of objects was presented on trial 1, then one of these objects was replaced by another on the remaining trials. For four monkeys, the object replaced was the one that they had chosen on trial 1; for the other four monkeys, the object that they had chosen was the one that was retained and paired with a new object. Objects that had been pushed aside always kept the same reward value, and if such an object were replaced by a new one on trial 2, the new object took on that reward value. Thus, to respond correctly on trial 2, monkeys in one group had to select the same object if the first-trial response had been rewarded, but had to select the new object if the first-trial response had been nonrewarded; these monkeys made about 80% correct responses on trial 2. The other group had to select the new object if the first-trial response had been rewarded, but had to select the object that had been present but not pushed aside if the first-trial response had been nonrewarded; these monkeys made about 50% correct responses on trial 2. In other words, the first-trial outcome was an effective cue only if the object that had been pushed aside was presented on the second trial. Both groups performed at high levels on trials 3 through 6.

The results of an earlier experiment by Leary (1956) similarly suggested that prior experience with the negative member of a stimulus pair facilitates discrimination performance only when the subject has actually displaced the negative object. The rather complicated procedure involved presenting a single object (either positive or negative) for two trials, pairing this object with a second object (previously positive with new negative or previously negative with new positive) for six trials, appropriately pairing the second object with a third object for six trials, and then, on half of the problems, appropriately pairing the third object with a fourth object for six more trials. Leary predicted that whenever the previously-used object in a newly-created pair was positive, the subject would rarely pick the new negative object; thus the effect of previous experience with a negative but unchosen object could be measured on subsequent trials. High positive transfer occurred when a previously displaced positive or negative object was paired with a new object, i.e., subjects selected a previously positive object or selected the new object paired with a previously chosen negative object. But when a previously *unchosen* negative object was paired with a new object, the subjects were inclined to select the negative (familiar) object rather than the positive (new) member of the stimulus pair. Later experiments in which novelty and familiarity served as cues for choice are described in Section V, B, 3.

5. DIFFERENTIAL EFFECTS OF REWARD AND NONREWARD

Moss and Harlow (1947) developed a procedure specifically designed to determine the relative effects of the two outcomes, reward and nonreward, upon discrimination efficiency on subsequent trials. Each problem consisted of 1 or 2 training trials followed by 11 discrimination test trials. Seven rhesus monkeys and one mangabey, with prior learning-set training, were given 30 discrimination problems under each of two training conditions: (1) The object designated positive for that problem was presented alone, covering food. (2) The negative object for that problem was presented alone, covering an empty foodwell. Thus, discrimination performance on test trials preceded by consistent reward of responses to the positive object could be compared with performance on trials preceded by consistent nonreward of responses to the negative object. The subjects showed highly efficient discrimination learning under either condition, but their performance on the first test trial was much better (about 95% versus about 75% correct responses) when training consisted of nonrewarded rather than rewarded object-displacement.

The same differential effect of nonreward and reward has been found by other investigators who used similar procedures. For example, eight rhesus monkeys were tested by Harlow and Hicks (1957) on 360 six-trial problems, in 180 of which they displaced a single object (positive in 90 problems, negative in the other 90) on trial 1 and then chose between this object and a new object on five discrimination-training trials. In a manner typical of learning-set formation, discrimination efficiency on trials 2 through 6 increased progressively. As in Moss and Harlow's experiment, discrimination performance following nonrewarded single-object displacement was consistently better than performance following preliminary reward (Fig. 14). Harlow and Hicks attributed this difference to the rhesus macaque's tendency to investigate, or "try out," a stimulus object which has not yet been explored ("response-shift" tendency; see Section IV, C, 1, *c*). When a subject displaces a single object during training, finds no food, and thereby learns to select the positive object on the first test trial, both objects have been investigated without penalty of error (i.e., response-shift tendency does not produce errors). On the other hand, a correct response following displacement of a single object covering food during training requires selection of the same object; the subject must commit at least one error on discrimination test trials in order to investigate the other object (i.e., response-shift tendency does produce errors).

King and Harlow (1962) also presented a single object on the first trial of each problem. But instead of rewarding responses to the single object in 50% of the problems, they modified this procedure for two of their three groups of four rhesus monkeys. The single object covered food in either 75% (group 75), 50% (group 50), or 25% (group 25) of 720 independent 2-trial problems. If the single object covered food on trial 1,

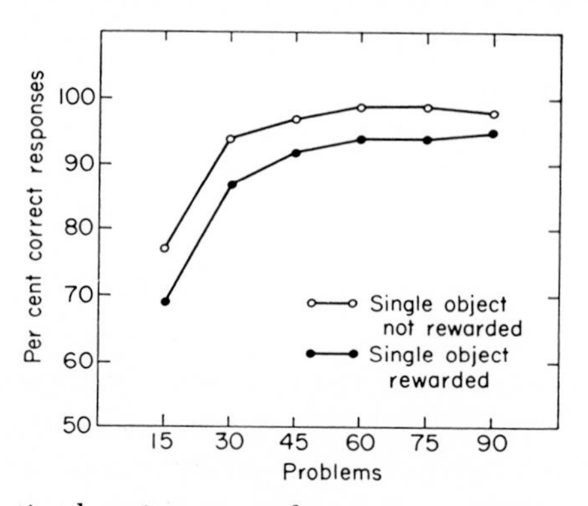

FIG. 14. Discrimination-learning-set performance on trials 2 through 6 following rewarded and nonrewarded responses on trial 1. (After Harlow & Hicks, 1957.)

this object was positive when paired with a new object on trial 2. If the response on trial 1 was nonrewarded, the new object was positive on trial 2. In accord with previous findings, second-trial performance of subjects in group 50 (equal reward and nonreward experience) improved progressively under both conditions, with consistently superior performance on discriminations preceded by nonreward. Of primary interest was the fact that subjects in groups 75 and 25 developed a general response pattern consistent with the relative frequency of preliminary reward and nonreward. Performance of group 75 improved markedly after preliminary reward but failed to improve after the less frequent event of nonreward. In group 25, performance after preliminary nonreward (which occurred in 75% of the problems) improved rapidly, while performance after the less frequent event of reward remained below chance. These data suggest that the subjects developed a response set according to the most frequent first-trial outcome. King and Harlow attributed these findings to the differential effect of each first-trial condition on error-producing response tendencies. The greater relative frequency of first-trial reward events for group 75 reversed the usually-found superiority of performance following nonreward by suppressing the response-shift tendency and thereby increasing stimulus-perseveration errors under the nonreward condition. In group 25, the stimulus-perseveration tendency was suppressed by the relative predominance of preliminary nonreward, resulting in more response-shift errors under the reward condition. In other words, when reward and nonreward did not occur equally often on trial 1, responses on trial 2 were controlled less by the properties of having been positive or negative and more by the properties of having been responded to or not responded to. Such properties can serve as cues as well as can outcomes of response. Experimental demonstrations of this are described next.

B. Other Cues for Interproblem Learning

1. Response-Shift-Learning Set

When reward on trial 2 depends on using the first-trial outcome as a cue, the tendency of rhesus monkeys to shift their responses to the object not previously chosen is an "error factor" to be eliminated. But reward of such "response-shift" behavior in an experiment by Brown and McDowell (1963) resulted in the development of a response-shift-learning set, in which the first-trial selection itself, rather than its outcome, was the cue for second-trial choice. Fifteen sophisticated male rhesus macaques were given 960 two-trial object-quality-discrimination problems. Pushing aside either object on the first trial procured food reward; on the second trial, food could be obtained only by "shifting" the response and pushing aside the object not selected on trial 1. Errors on trial 2 decreased from about 70% on the first 120 problems to about 10% at the end of the 960-problem series, indicating that the subjects had formed a response-shift-learning set. Brown and McDowell noted a contrast between this result and the finding of little, if any, interproblem improvement in the second-trial performance of three naive rhesus monkeys tested with a similar procedure by Riopelle and Francisco (1955). These subjects were tested on a series of 6-trial problems in which first-trial choice was invariably rewarded and selection of the other member of the stimulus pair was rewarded on trials 2 through 6. Under these conditions, subjects showed virtually no interproblem increase in proficiency (decrease in errors on trial 2) during 250 problems (see Chapter 5 by French).

The response-shift-learning-set procedure is analogous to a discrimination-reversal procedure in requiring the subject to shift an object-selection on the basis of a consistent cue. In reversal learning the cue is an outcome; in response-shift learning it is a previous response. But under Riopelle and Francisco's procedure, the subject must learn both to respond on trial 2 on the basis of the first-trial choice and to respond on trials 3 through 6 of the same problem on the basis of outcomes. Although monkeys can develop simultaneous sets to respond in conflicting ways to outcome cues, and can respond in nonconflicting ways to different types of cues (Section V, C), there is no evidence that they can learn to respond in conflicting ways to different types of cues, as Riopelle and Francisco's procedure required.

A response-shift-learning set could develop either through learning to "inhibit" response to the object previously chosen or through learning to displace the object not responded to on the first trial. To test these alternatives, McDowell and Brown (1963) first trained 18 sophisticated rhesus subjects on 1,056 two-trial response-shift problems (choice of either object rewarded on first trial, displacement of alternative object rewarded on second trial). Subjects then received 96 additional two-trial problems under each of the following conditions: Procedure A—The object responded to on the first trial was paired with a novel stimulus object on the second trial, and response to the novel object was rewarded. Procedure B—The object not responded to on the first trial was paired with a novel stimulus object on the second trial, and response to the object not previously responded to was rewarded. The established level of response-shift proficiency was not noticeably disturbed by Procedure A; subjects selected the novel object on the second trial, indicating that they had learned (through response-shift-learning-set training) to inhibit a previously rewarded response. But errors increased markedly under Procedure B, suggesting that the monkeys were learning very little about the object that they did not push aside on the initial trial (see Section V, A, 4).

2. Sequential-Order-Learning Set

Massar and Davis (1959) trained 13 experienced rhesus monkeys to solve discrimination problems by relying solely on cues obtained from the sequence in which objects were presented. Each independent problem consisted of a list of four stimulus objects arranged in a fixed serial order. One problem was presented on each of 63 test days. Preliminary training at the beginning of each problem consisted of repeated successive presentations of the four objects assigned to that problem: The first object was presented alone (in the center of the test tray) on the first trial, and displacing the object procured food; then the second, third, and fourth objects were presented according to the same procedure. Since the objects were always presented in the same sequential order, it was possible on later presentations of the list for the subject to "predict" the object that would be presented next. Subjects then received 12 consecutive repetitions of the list of objects under the experimental condition: On three of the four trials that constituted each repetition, the designated object was presented in the manner described for preliminary training. On the remaining trial, the critical experimental trial, two stimulus objects were presented (one at the right and one at the left of the test tray)—the positive member of the pair was the object that was in correct sequential order for that trial; the negative alternative was one of the three other objects from the same list (i.e., an object that was out of sequence on that trial). Better-than-chance performance on the critical trial thus indicated that the temporal sequence had been learned. Although improvement was slow and performance was still quite poor (about 40% errors) at the end of the experiment, frequency of errors on the critical trial decreased reliably as a function of practice on successive problems. Massar and Davis described this improvement in temporal-sequence learning as the formation of a temporal-sequence-learning set.

3. "NOVELTY" AND "FAMILIARITY"

Brown *et al.* (1958, 1959) demonstrated that rhesus monkeys can learn to respond to the abstract stimulus property of novelty. The procedure differed from conventional discrimination-learning-set training in that each stimulus pair was composed of one new object and one object retained from the immediately previous problem. In the 1958 study, the reward value of the object brought forward from the previous problem was reversed to assure that a problem could not be solved by making a learned response to a particular stimulus object. Forty-seven experienced rhesus monkeys were given 192 four-trial problems in which the negative stimulus object was retained to become positive when paired with a novel object in the next problem. Problems in which the positive object was brought forward and made negative when paired with a novel object were presented to eight relatively inexperienced subjects (240 problems) and five experienced subjects (96 problems). All groups showed a reliable decrease in number of first-trial errors as training progressed. Brown *et al.* (1958) claimed that the results showed the formation of a concept of "novelty" and the association of a response with the conceptual stimulus class. But their subjects could have learned either to reverse their object-selection when the novel stimulus object appeared as a cue or to respond four times to a given object and then reverse. So Brown *et al.* (1959) used a procedure designed to exclude such "simple" reversal solutions. They gave 33 experienced rhesus monkeys 192 four-trial problems in which one of the stimulus objects was retained from each problem and brought forward to be paired with a novel object on the next problem. In 96 problems the familiar object had been the positive member of the previous stimulus pair; in the other 96 it previously had been negative; and in all problems the novel object was positive, while the familiar object (regardless of previous reward value) became the negative member of the new stimulus pair. Thus, to show progressive improvement in first-trial performance, subjects had either to learn a conditional reversal (to reverse when positive object was retained, not to reverse when negative object was retained) or to learn to always choose the novel object. First-trial error scores under both conditions decreased steadily and reliably throughout the 192-problem series. A reliable difference between over-all performances under the two conditions was fairly constant throughout training, with about 30% errors when the positive object was brought forward and 15% errors when the negative object was brought forward during the last 24 problems. After discussing alternative explanations, Brown *et al.* (1959) reached the same conclusion as in the earlier study, that the results indicate the formation of a "novelty" concept.

An experiment by Riopelle *et al.* (1962) was directly concerned with interproblem effects of stimulus "familiarity" on discrimination learning. Twelve rhesus monkeys were given 10 six-trial problems each day for 60 days; four of these were "new" problems in which both stimulus objects were novel; the other six were "familiar" problems in which one member of each stimulus pair was retained from day to day. Four monkeys were tested under each of the following "familiarity" conditions: Group P—each of the six recurrent objects was the positive object in the problem in which it occurred. Group N—the six recurrent objects were always negative, and food was always placed under the novel member of the stimulus pair. Group P-N—the six recurrent objects appeared sometimes as the positive object and sometimes as the negative; reward value of a particular object was consistent within a problem but changed unpredictably from day to day. As would be expected, subjects learned to respond appropriately to familiar objects that retained a consistent reward value, either positive or negative, from day to day; first-trial performance of group P and of group N was reliably above chance, and discrimination efficiency rapidly increased to 100% correct responses. The performance of these two groups on "new" problems was differentially affected by the "familiarity" conditions: on the second trial of a "new" problem, monkeys in group P tended to repeat an initial correct choice, whereas monkeys in group N shifted response after a correct choice on the first trial. Thus, discrimination performance of group P benefited from experience with familiar stimuli in a consistently positive role. Group P-N, for which reward value of the familiar stimulus was unpredictable, necessarily performed at chance level on the first trial of "familiar" problems; their second-trial performance was worse than that of the other two groups on "familiar" problems and fell between the other groups on "new" problems. After this phase of the experiment was completed, all subjects were tested on 40 conventional discrimination-reversal problems. The subjects from group P-N, whose necessarily chance performance on the first trial during discrimination training had provided the most experience in shifting response from one object to the other, demonstrated clear superiority in discrimination-reversal learning.

Cross *et al.* (1963) also investigated the effects of stimulus familiarity on discrimination learning, but their familiarization technique consisted of giving subjects home-cage experience (rather than test experience) with objects to be used in discrimination problems. Sixteen rhesus monkeys received three 12-trial object-quality-discrimination problems each day for 50 days. Each of four groups of four monkeys was given a different home-cage condition during the 23.5 hours before a test session: For group P, the three objects that were to be positive in the three discrimination problems were placed in the home cage; group N was given the three objects that were to be negative; group I was given three objects that were to be either positive or negative (or not appear at all); and group C received no home-cage experience with stimulus objects. Groups P and N (consistent home-cage experience with either positive or negative test objects) learned to use "familiarity" cues in problem solution. First-trial performance of these groups improved throughout the course of the experiment, and they performed reliably better than did groups I and C. Comparisons between the performances of group P and group N showed reliably more correct first-trial responses by group N (rewarded for selecting the novel object), whereas the performance of group P (rewarded for choosing the familiar object) was reliably better on trials 2 through 12. The first-trial difference was explained in terms of a tendency to explore novel stimuli, and the difference on later trials was attributed to the response-shift tendency (Section IV, C, 1, *c*), which results in more errors after correct than after incorrect first-trial responses.

4. COLOR OF BACKGROUND

Most interpretations of repeated-reversal learning and of discrimination-reversal-learning set emphasize the "informative" (cue) function of nonreward. But several investigators have postulated that nonreward of a previously rewarded response produces an aversive emotional state, referred to as "frustration," which is supposed to have both motivational and cue properties. Nonreward of a previously rewarded response is an integral part of both simple reversal and the usual reversal-learning-set procedures, as the subject must make at least one nonrewarded response to the object that was positive on prereversal trials in order to be informed of the shift in reward value. It is possible, therefore, that induced "frustration" has had a significant effect upon discrimination-reversal-learning-set functions.

To determine if "frustration" plays a crucial role in discrimination-reversal learning, Riopelle and Copelan (1954) tested rhesus monkeys with a reversal procedure that did not require the subject to make a nonrewarded response. A change in the color of the test tray signified that reward values of paired stimulus objects had been switched, making it possible for the subject to reverse its object-selections without experiencing "frustrative" nonreward. Five subjects, which had been tested on series of discrimination and discrimination-reversal problems, each received 220 problems according to the following procedure: a pair of stimulus objects was presented on a green test tray for either six or eight prereversal trials; a yellow tray was then substituted and food was placed only under the previously negative object for six postreversal trials. By the end of the experiment, subjects were making almost 100% correct responses on the first postreversal trial, a result which clearly indicates that rhesus monkeys can become highly proficient at performing discrimination reversals to a discriminative cue (change in tray color) that does not involve "frustrative" nonreward. The discrimination-reversal-learning-set procedure used by Riopelle and Copelan is closely related to one usually referred to as "conditional discrimination" (see Chapter 5 by French). A change in background color signified that a previously positive object no longer covered food; thus, reward value of a given member of a stimulus pair was conditional on the color of the tray.

C. Simultaneous Development of Two Response Sets

Riopelle and Chinn (1961) required four rhesus monkeys to make a spatial discrimination when a novel pair of objects appeared, and then to respond to nonspatial cues on the basis of the first-trial outcome. The positive object for each problem always appeared on one side (in the left position for two subjects, right position for the other two) on the first trial; on trials 2 through 6 the same positive object shifted unpredictably from side to side, as in the usual object-quality-discrimination procedure. Figure 15 illustrates the progressive interproblem improvement in performance on trial 1 (spatial discrimination) and on trial 2 (object-quality discrimination) until, at the end of the 350-problem series, subjects were making about 85% responses to the appropriate position on trial 1 and to the correct object on trial 2. The high level of proficiency attained on both types of discrimination, and the fact that rewarded versus nonrewarded responses on trial 1 had no differential effect on performance on trial 2, suggest considerable independence of spatial and nonspatial discrimination performance. A consistent interproblem event (the appearance of a new stimulus pair) apparently became a cue for response to position, and the outcome of object selection on trial 1 evidently provided an intraproblem cue for selection of the positive object on trial 2.

In an earlier experiment by Zable and Harlow (1946), outcomes were both intraproblem and interproblem cues. Three sophisticated rhesus macaques learned to solve both spatial and nonspatial discrimination problems presented in random order. The monkeys first were trained on 60 object-quality, 30 left-position, and 30 right-position discriminations, each problem using a new pair of objects placed in right and left positions in irregular order. Then a single pair of objects was presented for 30 test sessions according to four different procedures: object-quality dis-

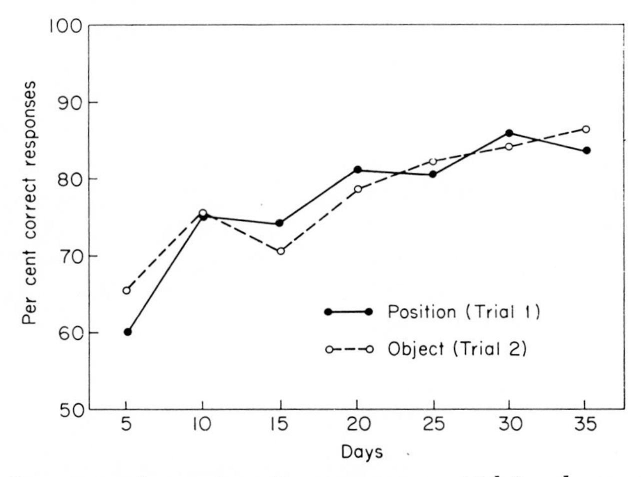

FIG. 15. Percentage of correct position responses on trial 1 and correct object responses on trial 2. (After Riopelle & Chinn, 1961.)

crimination (reward of responses to one object), object-discrimination reversal (reward of responses to the other object), position discrimination (reward of responses to one position), and position-discrimination reversal (reward of responses to the other position). Each of the four conditions was in effect for 15 consecutive trials during each session, with conditions arranged so that changes from one condition to another were unpredictable. The subject's task was to persist in some mode of behavior as long as reward was forthcoming and, when consistent reward no longer occurred, to try out different learned behavioral repertoires until it "discovered" the condition currently in effect. Not only did the monkeys learn to solve four different discrimination problems, involving contradictory behavior to both objects and positions, but they learned to shift from an inappropriate to an appropriate mode of behavior even though nonreward was the only cue signifying a change in conditions. The subjects' proficiency on position discriminations increased as the experiment progressed.

VI. THEORETICAL INTERPRETATIONS

A. "Uniprocess" Theory

Most learning theorists assume dual behavioral processes—"excitation" (reinforcement through reward) and "inhibition" (extinction through nonreward). Harlow (1959) feels that the systematic error trends exhibited in primate learning would be rather improbable if one assumed differential interactions between "excitatory" and "inhibitory" processes, and believes that they are more adequately thought of as representing the gradual elimination of inappropriate response tendencies. The results of an investigation by Harlow and Hicks (1957), which was summarized in Section V, A, 5, are taken as fairly strong support for this "uniprocess" learning theory. The two learning-set curves shown in Fig. 14, representing performance following rewarded and nonrewarded first-trial responses, are parallel and alike in form throughout learning-set development. Harlow and Hicks interpreted the similarity in form of the two functions as reflecting a single basic learning process, since "differential curve form . . . would result if reward and nonreward showed differential strengthening and weakening effects during the course of learning" (Harlow, 1959, p. 527).

Description of learning-set development as the suppression of error factors (Section III, C, 1) resulted from the arbitrary decision to analyze discrimination errors rather than correct responses. A progressive interproblem increase in correct responses is, of course, the over-all trend, and suppression of error-producing behavior is always correlated with an increase in correct choices. In his original error-factor article, Harlow stated, "It would be equally justifiable to describe the learning in terms of positive response tendencies" (Harlow, 1950a, p. 37). More recently, he has concluded that the one basic process of learning is inhibition, and that learning-set formation *is* the gradual suppression of various inappropriate modes of behavior (Harlow, 1959). The correct response is assumed to be already a part of the animal's behavioral repertoire, but it is "blocked" by many inappropriate responses to the learning situation. Through learning-set training, these error-producing modes of responding are weakened or eliminated, allowing the correct response to occur without interference. Thus, the correct response is not considered to be reinforced or changed in any way through reward, but remains the activity that the subject exhibits because the probabilities of all other behaviors decrease.

The further development of this "uniprocess" theory, which is still in its formative stages, would require a more formal statement of some clearly specified assumptions or postulates. It would then be possible to evaluate the theory empirically and to determine to what extent error-suppression is a general principle that can be applied to learning situations other than discrimination-learning-set formation.

Although Harlow's "uniprocess" theory of learning is compatible with the error-factor data from which it originated, the discrimination-learning-set performance of rhesus macaques is not completely represented by the error factors that have already been identified. Further analyses will presumably reveal other response tendencies correlated with incorrect object-selection. Selection of a particular stimulus object undoubtedly is a complex multidetermined event. For example, displacement of an object on the first trial of a discrimination problem could result from interaction of various factors, such as location of the subject near that object when the opaque screen is raised, responding to the position where food was procured on the last trial(s) of the preceding problem, comparable avoidance of the position where food was not obtained, "innate" attraction to the object, "innate" fear of the alternative object, learned preference for the object because of similarity to a previously positive object, and avoidance of the alternative object because of similarity to a previously negative object. The reward or nonreward outcome of this initial choice is highly relevant to object-selection on trial 2, but all of the foregoing variables could continue to influence performance.

B. Restle's Mathematical Model

There are two quantitative interpretations of discrimination-learning-set formation (Levine, 1959; Restle, 1958). Because Levine's formulation is discussed at length in Chapter 3, only Restle's approach will be presented here.

Restle's (1958) quantitative account of learning-set formation is a further development of his earlier theory of single-problem discrimination and choice learning (Restle, 1955, 1957) and includes simple discrimination behavior as a special case. The theory assumes that three classes of cues are present during the formation of a discrimination-learning set. "Type-*a*" cues are those cues which are valid (consistently correlated with reward) after the first, "informative" trial, and common to all of the independent discrimination problems. These "type-*a*" cues are considered to be "abstract" (because the subject's response becomes independent of concrete properties of particular stimulus objects) and also "conditional" (because object choice is contingent on the outcomes of previous object-selections). In other words, the subject "is responding to the property of having been reinforced, as distinguished from other properties such as size, color, spatial shape, arrangement of parts, etc." (Restle, 1958, p. 79). Concrete properties of stimulus objects (size, color, etc.) give rise to "type-*b*" cues, which are valid (correlated with reward) within each problem but not from problem to problem. Cues of a third type are "invalid" (not consistently correlated with reward at any time); these "type-*c*" cues may originate from the apparatus, from within the animal, and from procedural events uncorrelated with reward.

In general, Restle's theory assumes that the development of a learning set occurs as the subject learns to utilize type-*a* cues and to "ignore" cues that are invalid (type-*c*) and those valid only within individual problems (type-*b*). Cues that are ignored, or "adapted," have no effect on the probability of a correct response. This probability, on any given trial, is the proportion of unadapted cues that are conditioned to the correct response. The rate at which cues are conditioned depends on the mean validity of the total set of available cues. The valid cues become conditioned to the correct response through consistent reinforcement. The invalid cues, on the other hand, become conditioned to either response. Invalid cues eventually become adapted, but until the process is complete they reduce the proportion of valid cues, thus reducing the probability of the correct response. On any given discrimination trial, both processes occur—conditioning of a constant proportion of all cues, and adaptation of a constant proportion of invalid cues. It is assumed that invalid cues become adapted with respect to currently present valid cues and remain adapted only in the presence of these valid cues. Those type-*c* cues that adapt with respect to problem-specific, type-*b* cues must be readapted on following problems containing new type-*b* cues. But when a type-*c* cue becomes adapted with respect to a type-*a* cue, the effect is permanent, since, by definition, type-*a* cues are common to all problems. The type-*b* cues have temporary intraproblem validity, but are not consistently correlated with reinforcement from problem to problem and slowly become adapted with respect to the consistently valid type-*a* cues. As this occurs, the proportion of unadapted cues that are type-*a* cues approaches 1.00, and instantaneous solution of discrimination problems following the first "informative" trial becomes possible. The general description of learning-set behavior that follows from these assumptions is in accord with empirical phenomena: the theory predicts a gradual and continuous interproblem improvement in performance culminating in the one-trial solution of independent discrimination problems.

From these assumptions, Restle was able to derive equations representing the conditioning of the three types of cues, the slow, permanent adaptation of type-*b* cues, and the temporary and permanent adaptation of type-*c* cues. These equations were then combined into a general formula expressing the probability of a correct response on any given trial of a learning-set experiment. In order to test the theory, the three main independent parameters (proportion of type-*a* cues, proportion of type-*b* cues, and the validity of type-*b* cues) were estimated by a method of approximations from learning-set data reported by Harlow (1949). With these estimated parameters, the general formula produced theoretical learning functions which corresponded quite closely to Harlow's eight empirical intraproblem learning curves shown in Fig. 1. A slightly modified version of the original formula enabled Restle to predict, with a fair degree of accuracy, the performance of sophisticated subjects on discrimination-reversal problems (Meyer, 1951a, 1951b; see Section II, B) and on pattern discriminations under a variety of differential-cue conditions in Warren's (1953) experiment on additivity of cues (see Chapter 1 by Meyer *et al.*).

C. A Hull-Spence Approach

After a selective review of the literature on discrimination-learning-set formation, Reese (1964) proposed that one-trial learning could be accounted for by the Hull-Spence theory. According to this approach, one-trial learning could occur if "(*a*) an increment in habit strength produced by a single reinforcement or an increment in inhibitory strength produced by a single nonreinforcement makes the excitatory potential of the positive stimulus greater than that of the negative stimulus, and (*b*) the difference between the excitatory potentials of the positive and negative stimuli is greater than oscillatory inhibition" (Reese, 1964, p. 334). Long training is assumed to reduce oscillatory inhibition and to equalize the habit strengths and inhibitory strengths of new stimuli before they are responded to. The equalization is supposed to occur through roughly equal frequencies of reward and nonreward of responses to large numbers of stimulus elements and through generalization to unused elements. Thus, "the theory requires that the subject be given practice on a large number of problems" (Reese, 1964, p. 335) involving many different stimulus elements.

Most learning-set training procedures meet the requirements of this theory, but two of the experiments described in Section V, A, 3 apparently do not. Riopelle's (1955) experiment on "learning sets from minimal stimuli" used four stimulus objects, and Schusterman (1962) used three pairs of objects to produce a discrimination-learning set by repeated-reversal training. It does not seem reasonable to posit that four or six test objects would contain enough stimulus elements to equalize habit strengths and inhibitory strengths of new objects. It also is difficult to visualize how the theory could be expanded to account for data from some of the special procedures such as object-alternation-learning set (Behar, 1961; see Chapter 5 by French and Chapter 3 by Levine).

Although the theoretical interpretations which analyze hypotheses, strategies, or error factors can handle rather diversified data, there is as yet no clear explanation for the strengthening or weakening of these response tendencies (Reese, 1964). It seems reasonable to predict, however, that the current proliferation of learning-set research will soon produce a theoretical blend which will both handle a variety of empirical phenomena and provide a precise account of the development of underlying response sets.

REFERENCES

Behar, I. (1961). Analysis of object-alternation learning in rhesus monkeys. *J. comp. physiol. Psychol.* **54,** 539.

Bowman, R. E. (1963). Discrimination learning-set performance under intermittent and secondary reinforcement. *J. comp. physiol. Psychol.* **56,** 429.

Braun, H. W., Patton, R. A., & Barnes, H. W. (1952). Effects of electro-shock convulsions upon the learning performance of monkeys: I. Object-quality discrimination learning. *J. comp. physiol. Psychol.* **45,** 231.

Brown, W. L., & McDowell, A. A. (1963). Response shift learning set in rhesus monkeys. *J. comp. physiol. Psychol.* **56,** 335.

Brown, W. L., Overall, J. E., & Gentry, G. V. (1958). Conceptual discrimination in rhesus monkeys. *J. comp. physiol. Psychol.* **51,** 701.

Brown, W. L., Overall, J. E., & Blodgett, H. C. (1959). Novelty learning sets in rhesus monkeys. *J. comp. physiol. Psychol.* **52,** 330.

Cross, H. A., Fletcher, H. J., & Harlow, H. F. (1963). Effects of prior experience with test stimuli on learning-set performance of monkeys. *J. comp. physiol. Psychol.* **56,** 204.

Darby, C. L., & Riopelle, A. J. (1955). Differential problem sequences and the formation of learning sets. *J. Psychol.* **39,** 105.

Darby, C. L., & Riopelle, A. J. (1959). Observational learning in the rhesus monkey. *J. comp. physiol. Psychol.* **52,** 94.

Grandine, Lois, & Harlow, H. F. (1948). Generalization of the characteristics of a single learned stimulus by monkeys. *J. comp. physiol. Psychol.* **41,** 327.

Harlow, H. F. (1944). Studies in discrimination learning by monkeys: I. The learning of discrimination series and the reversal of discrimination series. *J. gen. Psychol.* **30,** 3.

Harlow, H. F. (1949). The formation of learning sets. *Psychol. Rev.* **56,** 51.

Harlow, H. F. (1950a). Analysis of discrimination learning by monkeys. *J. exp. Psychol.* **40,** 26.

Harlow, H. F. (1950b). Performance of catarrhine monkeys on a series of discrimination reversal problems. *J. comp. physiol. Psychol.* **43,** 231.

Harlow, H. F. (1959). Learning set and error factor theory. *In* "Psychology: A Study of a Science" (S. Koch, ed.), Vol. 2, pp. 492-537. McGraw-Hill, New York.

Harlow, H. F., & Hicks, L. H. (1957). Discrimination learning theory: uniprocess vs. duoprocess. *Psychol. Rev.* **64,** 104.

Harlow, H. F. & Poch, Susanne (1945). Discrimination generalization by macaque monkeys to unidimensional and multidimensional stimuli. *J. comp. Psychol.* **38,** 353.

Hayes, K. J., Thompson, R., & Hayes, Catherine (1953a). Discrimination learning sets in chimpanzees. *J. comp. physiol. Psychol.* **46,** 99.

Hayes, K. J., Thompson, R., & Hayes, Catherine (1953b). Concurrent discrimination learning in chimpanzees. *J. comp. physiol. Psychol.* **46,** 105.

Jackson, T. A. (1932). General factors in transfer of training in the white rat. *Genet. Psychol. Monogr.* **11**, 3.

King, J. E., & Harlow, H. F. (1962). Effect of ratio of trial 1 reward to nonreward on the discrimination learning of macaque monkeys. *J. comp. physiol. Psychol.* **55**, 872.

Lashley, K. S., & Wade, Marjorie (1946). The Pavlovian theory of generalization. *Psychol. Rev.* **53**, 72.

Leary, R. W. (1956). The rewarded, the unrewarded, the chosen, and the unchosen. *Psychol. Rep.* **2**, 91.

Leary, R. W. (1957). The effect of shuffled pairs on the learning of serial discrimination problems by monkeys. *J. comp. physiol. Psychol.* **50**, 581.

Leary, R. W. (1958a). Analysis of serial-discrimination learning by monkeys. *J. comp. physiol. Psychol.* **51**, 82.

Leary, R. W. (1958b). Homogeneous and heterogeneous reward of monkeys. *J. comp. physiol. Psychol.* **51**, 706.

Leary, R. W. (1958c). The temporal factor in reward and nonreward of monkeys. *J. exp. Psychol.* **56**, 294.

Leary, R. W. (1962). "Spontaneous reversal" in serial-discrimination reversal learning of monkeys. *Canad. J. Psychol.* **16**, 228.

Levine, M. (1959). A model of hypothesis behavior in discrimination learning set. *Psychol. Rev.* **66**, 353.

Levine, M., Levinson, Billey, & Harlow, H. F. (1959). Trials per problem as a variable in the acquisition of discrimination learning set. *J. comp. physiol. Psychol.* **52**, 396.

Levinson, Billey (1958). Oddity learning set and its relation to discrimination learning set. Doctoral dissertation, University of Wisconsin. University Microfilms, Ann Arbor, Michigan, No. 58-7507.

Lockhart, J. M., Parks, T. E., & Davenport, J. W. (1963). Information acquired in one trial by learning-set experienced monkeys. *J. comp. physiol. Psychol.* **56**, 1035.

McDowell, A. A., & Brown, W. L. (1963). The learning mechanism in response shift learning set. *J. comp. physiol. Psychol.* **56**, 572.

Maier, N. R. F., & Schneirla, T. C. (1935). "Principles of Animal Psychology." McGraw-Hill, New York.

Marx, M. H. (1944). The effects of cumulative training upon retroactive inhibition and transfer. *Comp. Psychol. Monogr.* **18**, No. 2 (Serial No. 94).

Massar, R. S., & Davis, R. T. (1959). The formation of a temporal-sequence learning set by monkeys. *J. comp. physiol. Psychol.* **52**, 225.

Melton, A. W., & Von Lackum, W. J. (1941). Retroactive and proactive inhibition in retention: evidence for a two-factor theory of retroactive inhibition. *Amer. J. Psychol.* **54**, 157.

Menzel, E. W., Jr. (1962). The effects of stimulus size and proximity upon avoidance of complex objects in rhesus monkeys. *J. comp. physiol. Psychol.* **55**, 1044.

Meyer, D. R. (1951a). Food deprivation and discrimination reversal learning by monkeys. *J. exp. Psychol.* **41**, 10.

Meyer, D. R. (1951b). Intraproblem-interproblem relationships in learning by monkeys. *J. comp. physiol. Psychol.* **44**, 162.

Meyer, D. R. (1951c). The effects of differential rewards on discrimination reversal learning by monkeys. *J. exp. Psychol.* **41**, 268.

Miles, R. C. (1957). Learning-set formation in the squirrel monkey. *J. comp. physiol. Psychol.* **50**, 356.

Miles, R. C. (1959). Discrimination in the squirrel monkey as a function of deprivation and problem difficulty. *J. exp. Psychol.* **57**, 15.

Miles, R. C., & Meyer, D. R. (1956). Learning sets in marmosets. *J. comp. physiol. Psychol.* **49,** 219.

Miller, R. E. Murphy, J. V., & Finocchio, D. V. (1955). A consideration of the object-quality discrimination task as a dependent variable. *J. comp. physiol. Psychol.* **48,** 29.

Moss, Eileen M., & Harlow, H. F. (1947). The role of reward in discrimination learning in monkeys. *J. comp. physiol. Psychol.* **40,** 333.

Reese, H. W. (1964). Discrimination learning set in rhesus monkeys. *Psychol. Bull.* **61,** 321.

Restle, F. (1955). A theory of discrimination learning. *Psychol. Rev.* **62,** 11.

Restle, F. (1957). Theory of selective learning with probable reinforcements. *Psychol. Rev.* **64,** 182.

Restle, F. (1958). Toward a quantitative description of learning set data. *Psychol. Rev.* **65,** 77.

Riopelle, A. J. (1953). Transfer suppression and learning sets. *J. comp. physiol. Psychol.* **46,** 108.

Riopelle, A. J. (1955). Learning sets from minimum stimuli. *J. exp. Psychol.* **49,** 28.

Riopelle, A. J., & Chinn, R. McC. (1961). Position habits and discrimination learning by monkeys. *J. comp. physiol. Psychol.* **54,** 178.

Riopelle, A. J., & Churukian, G. A. (1958). The effect of varying the intertrial interval in discrimination learning by normal and brain-operated monkeys. *J. comp. physiol. Psychol.* **51,** 119.

Riopelle, A. J., & Copelan, E. L. (1954). Discrimination reversal to a sign. *J. exp. Psychol.* **48,** 143.

Riopelle, A. J., & Francisco, E. W. (1955). Discrimination learning performance under different first-trial procedures. *J. comp. physiol. Psychol.* **48,** 90.

Riopelle, A. J., Francisco, E. W., & Ades, H. W. (1954). Differential first-trial procedures and discrimination learning performance. *J. comp. physiol. Psychol.* **47,** 293.

Riopelle, A. J., Cronholm, J. N., & Addison, R. G. (1962). Stimulus familiarity and multiple discrimination learning. *J. comp. physiol. Psychol.* **55,** 274.

Rumbaugh, D. M., & McQueeney, J. A. (1963). Learning-set formation and discrimination reversal: Learning problems to criterion in the squirrel monkey. *J. comp. physiol. Psychol.* **56,** 435.

Rumbaugh, D. M., & Prim, Merle M. (1963). A comparison of learning-set training methods and discrimination reversal training methods with the squirrel monkey. *Amer. Psychologist* **18,** 408. (Abstract.)

Schrier, A. M. (1958). Comparison of two methods of investigating the effect of amount of reward on performance. *J. comp. physiol. Psychol.* **51,** 725.

Schrier, A. M., & Harlow, H. F. (1956). Effect of amount of incentive on discrimination learning by monkeys. *J. comp. physiol. Psychol.* **49,** 117.

Schusterman, R. J. (1962). Transfer effects of successive discrimination-reversal training in chimpanzees. *Science* **137,** 422.

Spence, K. W. (1936). The nature of discrimination learning in animals. *Psychol. Rev.* **43,** 427. (Reprinted in "Behavior Theory and Learning," pp. 269-291. Prentice-Hall, Englewood Cliffs, New Jersey, 1960.)

Sterritt, G. M., Goodenough, Eva, & Harlow, H. F. (1963). Learning set development: trials to criterion vs six trials per problem. *Psychol. Rep.* **13,** 267.

Strong, P. N., Jr. (1959). Memory for object discriminations in the rhesus monkey. *J. comp. physiol. Psychol.* **52,** 333.

Ward, L. B. (1937). Reminiscence and rote learning. *Psychol. Monogr.* **49,** No. 4 (Whole No. 220).

Warren, J. M. (1953). Additivity of cues in visual pattern discrimination by monkeys. *J. comp. physiol. Psychol.* **46,** 484.

Warren, J. M. (1954). Reversed discrimination as a function of the number of reinforcements during pre-training. *Amer. J. Psychol.* **67**, 720.

Wiltbank, R. T. (1919). Transfer of training in white rats upon various series of mazes. *Behav. Monogr.* **4,** No. 1 (Serial No. 17).

Zable, Myra, & Harlow, H. F. (1946). The performance of rhesus monkeys on series of object-quality and positional discriminations and discrimination reversals. *J. comp. Psychol.* **39,** 13.

Chapter 3

Hypothesis Behavior[1]

Marvin Levine

Department of Psychology, Indiana University, Bloomington, Indiana

I. INTRODUCTION

Systematic sequences of choice responses have been observed sporadically since the earliest experimental studies of primate learning. Hamilton (1911) recorded some sequential responding by macaque monkeys (*Macaca* spp.) in a four-choice apparatus. Yerkes (1916), using multiple-choice problems, noticed that an orangutan (*Pongo pygmaeus*) habitually tried first the alternative nearest the starting point. Gellermann (1933a) described position preferences and alternation patterns manifested by chimpanzees (*Pan*) during discrimination learning, and Spence (1937) described a stimulus preference.

Interest in these systematic patterns of response was increased both by the introduction of the object-discrimination-learning-set (ODLS)

[1] This chapter was written while I was a NATO fellow at the Laboratoire de Psychologie, Université Libre de Bruxelles, in Belgium. Deep gratitude is due the personnel of the laboratory for their assistance.

experiment (Harlow, 1949) and by Harlow's (1950) analysis of the resulting data (see Chapter 2 by Miles). The chief virtue of the learning-set procedure is that the short problem forms a convenient unit for measuring a pattern of response. Also, several hundred problems are typically presented before near-perfect performance is achieved. The researcher, therefore, has enough units (problems) to describe a variety of systematic behaviors at each stage of learning. Harlow (1950) analyzed the learning-set data for four different response patterns which he called "error factors" because they produce errors in the ODLS experiment. The four error factors were position habit (preponderance of response to one of the two sides), stimulus perseveration (relative excess of consecutive errors following a first-trial error), differential cue (response to the position previously yielding food rather than to the object), and response shift (excessive response to the negative object following response to the positive object). Impelled by this work, several investigators measured error factors during the decade that followed. Among the more unusual analyses were those by Moon and Harlow (1955), demonstrating error-factor functions during the formation of oddity-learning sets (see Chapter 5 by French), by Schrier and Harlow (1956), showing the influence of amount of reward on the response patterns, and by Leary (1958), applying the measures to concurrent-discrimination learning. A detailed review of several of these studies was presented by Harlow (1959; see also Chapter 2 by Miles).

This series of analyses contained an imperfection: the measures were separately improvised for each error factor. As a result, the proportion of responses controlled by a given error factor could not be determined, nor could the strengths of the various error factors be compared.

To eliminate this problem, Levine (1959b) developed a miniature theoretical framework for treating a large variety of systematic patterns of response in the ODLS experiment. His model had several special features. First, the foregoing difficulties were eliminated: all patterns were measured in a standard manner, yielding estimates of the proportion of responses which were manifestations of each tendency, and these proportions were additive. Second, the model was mathematical and permitted the derivation of theorems which could be tested experimentally. Third, a large number of patterns could be measured. Nine were considered and will be discussed below. Fourth, since learning also is manifested as a systematic pattern of response, the model was generalized to include both error-producing and reward-producing patterns. Rather than use the term "error factors," therefore, the term "hypotheses," used by Krechevsky (1932) to describe systematic response patterns of rats, was adopted.

The model was restricted to three-trial problems. For longer problems, only the first three trials of each were analyzed. The model's adequacy for explaining the distribution of response sequences was demonstrated for several ODLS series. This work will be summarized here, followed by further explorations of the same type of data, by an experiment designed specifically to test the model, and by an application of the model to oddity learning.

II. THE HYPOTHESIS MODEL AND ITS APPLICATION TO THE OBJECT-DISCRIMINATION-LEARNING-SET EXPERIMENT

A. The Object-Discrimination-Learning-Set Experiment

The phenomena of hypothesis behavior, the problems posed, and their solution will be better revealed if the procedure of the three-trial-per-problem ODLS experiment is described more abstractly. For each three-trial problem the positive object typically changes position in some previously determined but unsystematic manner. Placement of the positive object on the left (L) or right (R) may be designated, for example, LLR in the first problem, RRR in the second problem, etc. For any three-trial problem there are eight possible sequences. These may be paired into symmetrical pairs of (LLL, RRR), (LLR, RRL), (LRL, RLR), and (LRR, RLL). Initially, the discussion of hypothesis behavior will consider only the pattern of the sequence and will ignore the particular positions involved on any trial. There are, then, four sequences: AAA, AAB, ABA, and ABB, where A is defined as the location of reward on the first trial, and B is defined as the other position. These sequences will be referred to as Reward Sequences.

It is desirable to characterize responses in a similar manner. The Response Sequences may be described as III, II0, I0I, and I00, where I is the position responded to on the first trial and 0 is the other position. It will also be useful to describe the sequence of responses to the particular objects. A response to the positive object (or, equivalently, a response that is rewarded) will be characterized as +; a response to the negative object (not rewarded) will be characterized as —. The resulting sequences will be described at Outcome Sequences.

Knowledge of the Reward Sequence and of the Response Sequence does not define unequivocally the Outcome Sequence. For example, if the reward has gone AAA and the subject has responded III, one may conclude only that either +++ or ——— has occurred. There are, then, 4 Reward Sequences, 4 Response Sequences, and for each of the 16 pairings of these, 2 possible Outcome Sequences. In all, therefore, there are 32 different events which may occur in any problem.

Figure 1 presents a typical table of the frequencies of the 32 events.

These data[2] are from 100 problems presented to each of 10 rhesus monkeys (*Macaca mulatta*). The cells are arranged so that those Outcome Sequences ending in $-_3$ (response to the negative object on the third trial) are in the matrix on the left, those ending in $+_3$ are on the right. The frequencies appear to fall into some odd patterns. First, the frequencies in the right-hand matrix are almost uniformly larger than those in the left-hand matrix. This particular pattern is no surprise since this is a learning experiment and those patterns that are more apt to reveal learning, i.e., that end in $+_3$, have been grouped on the right. Second, the first, or left-hand, column in each of the two matrices has larger fre-

Reward sequence	Response sequence III	IIO	IOI	IOO
AAA	acfr 48 $-_1 -_2 -_3$	r 15 $+_1 +_2 -_3$	bdr 8 $-_1 +_2 -_3$	fr 12 $+_1 -_2 -_3$
AAB	aer 40 $+_1 +_2 -_3$	cr 42 $-_1 -_2 -_3$	br 15 $+_1 -_2 -_3$	der 18 $-_1 +_2 -_3$
ABA	adr 35 $-_1 +_2 -_3$	er 8 $+_1 -_2 -_3$	bcer 42 $-_1 -_2 -_3$	r 15 $+_1 +_2 -_3$
ABB	ar 22 $+_1 -_2 -_3$	dfr 9 $-_1 +_2 -_3$	bfr 15 $+_1 +_2 -_3$	cr 28 $-_1 -_2 -_3$

Reward sequence	Response sequence III	IIO	IOI	IOO
AAA	$acep_2r$ 100 $+_1 +_2 +_3$	p_3r 25 $-_1 -_2 +_3$	bdp_3r 12 $+_1 -_2 +_3$	ep_2r 30 $-_1 +_2 +_3$
AAB	afp_3r 40 $-_1 -_2 +_3$	cp_2r 60 $+_1 +_2 +_3$	bp_2r 25 $-_1 +_2 +_3$	dfp_3r 10 $+_1 -_2 +_3$
ABA	adp_3r 38 $+_1 -_2 +_3$	fp_2r 22 $-_1 +_2 +_3$	$bcfp_2r$ 60 $+_1 +_2 +_3$	p_3r 30 $-_1 -_2 +_3$
ABB	ap_2r 52 $-_1 +_2 +_3$	dep_3r 22 $+_1 -_2 +_3$	bep_3r 32 $-_1 -_2 +_3$	cp_2r 70 $+_1 +_2 +_3$

FIG. 1. A set of frequencies for each of the 32 combinations of Reward and Response Sequences.

quencies than the other three columns (the Response Sequence III occurs 375 times; each of the other three sequences occurs 203, 209, 213 times respectively). Third, in each matrix the diagonal from the upper-left- to the lower-right-hand corner contains relatively large frequencies. Other less obvious patterns may exist. Any adequate theory of systematic patterns of responding must take account of the variations which appear in a table of this sort.

B. Assumptions of the Hypothesis Model

(1) There is available to the subject a variety of hypotheses (*Hs*) among which he chooses at the outset of each problem. The *H* can be such tendencies as Position Preference, Stimulus Preference, a tendency to alternate positions or stimuli, or to respond in terms of outcomes on previous trials. The list of *Hs* assumed for the ODLS experiment will be presented formally below.

[2] Data from Harlow *et al.* (1960) were modified to eliminate irrelevant variations. The relationships described in the text, however, existed in the original data as well.

(2) One and only one *H* is selected at the outset of the problem. Thus, the *H*s are mutually exclusive. Their stipulation will be such that the set will also be exhaustive.

(3) The *H* selected at the outset of the problem determines the behavior during the three trials of that problem in a precisely specifiable way. Note the distinction between the *H* and the behavior: the *H* determines the behavior. That is, the *H* will be treated as a mediating process, whose relation to behavior is explicitly defined. A second feature of this assumption is that the behavior is determined for all three trials of the problem. This is equivalent to asserting that the subject does not change *H*s during a problem.

The above three assumptions are quite general and, except for changes in a few words in assumption (3) (necessitated by the restriction there to three-trial problems), could be made for an *H* analysis of the behavior of any species on any task consisting of a series of discrete trials with clearly-defined response classes. The next assumption, however, concerning the set of *H*s that are available, requires intimate acquaintance with the species, the experience of the subjects, and the task. Thus, different sets of *H*s would be postulated for adult humans (see Levine, 1963), for rats (see Pubols, 1962), for monkeys that had learned double alternation and for those that had not, etc. Assumption (4), then, is specific to experimentally naive monkeys in the ODLS experiment.

(4) Nine and only nine *H*s are assumed. For each one, the behavioral manifestation is described and, in parentheses, summarized.

H_a : Position Preference: responding to only one position (III).

H_b : Position Alternation: alternating between positions from trial to trial (I0I).

H_c : Stimulus Preference: responding to only one object (+++ or −−−).

H_d : Stimulus Alternation: alternating between objects from trial to trial (+−+ or −+−).

H_e : Win-stay-Lose-shift with respect to position: repeating response to a position after reward and changing response to the other position after nonreward (I+,I+,I or I+,I−,0 or I−,0+,0 or I−,0−,I).

H_f : Lose-stay-Win-shift with respect to position: changing response to the other position after reward and repeating response to a position after nonreward (I+,0+,I or I+,0−,0 or I−,I+,0 or I−,I−,I).

H_{p_2} : Win-stay-Lose-shift with respect to the object: repeating response to an object after reward and changing response to the other object after nonreward (+++ or −++). This *H* yields maximum reward in the learning-set situation, since correct responding begins on the second trial of the problem.

H_{p_3} : Third-trial Learning: correct response on the third trial although not on the second trial ($+-+$ or $--+$). This H is required by the well-known fact that the correct response may suddenly appear on later trials.[3]

H_R : The Residual category: It is assumed that there are other Hs occurring to unrecorded stimuli such as emotional changes, proprioceptive cues, extraneous noises, lights, etc. These are all combined and treated as a single H whose characteristic is that all sequences are equally likely ($+++$, $++-$, . . . , $---$).

(5) Associated with each H will be a variable denoting the probability of its occurrence. This may be interpreted as a relative frequency of occurrence of the H. The subscript symbol will also serve as the probability symbol. Thus: $P(H_a) = a$, $P(H_b) = b$, ... Assumptions (2) and (5) permit the statement:

$$a + b + c + d + e + f + p_2 + p_3 + R = 1.00. \tag{1}$$

C. Determination of the *H* Probabilities

The fundamental problem for the model is to numerically evaluate the H probabilities from a set of data. This problem is solved by determining first the theoretical probability of each sequence (represented by each cell) in Fig. 1. To take a specific instance, one must determine the probability of $+_1 +_2 +_3$ when AAA is presented. The technique for solution has been described by Levine (1959b). For this particular example,

$$P(+_1 +_2 +_3 \mid \text{AAA}) = P(+_1 \mid \text{AAA})\ [a + c + e + p_2 + r],$$

where $r = R/4$. This may be rewritten as

$$P(+_2 +_3 \mid \text{A}+_1\text{AA}) = a + c + e + p_2 + r. \tag{2}$$

Thus, the probability related to the sequence (AAA, $+_1 +_2 +_3$) may be expressed as a linear combination of certain H probabilities. A similar equation may be derived for each of the 32 possible sequences. For example,

$$P(-_2 +_3 \mid \text{A}-_1\text{AB}) = a + f + p_3 + r,$$
$$P(-_2 +_3 \mid \text{A}+_1\text{BA}) = a + d + p_3 + r,$$
etc.

The probability symbols appearing in each equation are shown in the corresponding cell of Fig. 1.

[3] This H, unlike the others, does not immediately appear to be a unitary process. The exact description of its manifestation, however, permits it to be treated as any other H without risk of inconsistency.

The probability of a sequence may also be estimated from frequencies obtained in the experiment. To continue with the example above,

$$P(+_2+_3 \mid \mathrm{A}+_1\mathrm{AA}) = \frac{n(\mathrm{A}+_1\mathrm{A}+_2\mathrm{A}+_3)}{n(\mathrm{A}+_1\mathrm{AA})}, \tag{3}$$

where $n(\mathrm{A}+_1\mathrm{A}+_2\mathrm{A}+_3)$ and $n(\mathrm{A}+_1\mathrm{AA})$ are the number of times that $\mathrm{A}+_1\mathrm{A}+_2\mathrm{A}+_3$ and $\mathrm{A}+_1\mathrm{AA}$, respectively, have occurred. From Eqs. (2) and (3) it follows that

$$a + c + e + p_2 + r = \frac{n(\mathrm{A}+_1\mathrm{A}+_2\mathrm{A}+_3)}{n(\mathrm{A}+_1\mathrm{AA})}.$$

In the hypothetical set of data presented in Fig. 1,

$$a + c + e + p_2 + r = \frac{100}{15 + 12 + 100 + 12} = 0.719.$$

This is a linear equation in five unknowns. A total of 32 such equations may be obtained.

It should be clear that a solution has been produced for the problem of evaluating the *H* probabilities in a set of data. Since there are nine variables to be determined one need merely select nine of the 32 equations and solve the system of simultaneous equations. This will be referred to as Method I. It offers the advantage that the estimates of the *H* probabilities obtained from the nine equations may be applied to the other 23 equations to determine the adequacy of the model for describing a block of data of the sort presented in Fig. 1. This advantage will be exploited later.

A second method (Method II) exists for determining the values of the nine unknowns. This is the classical method of least squares and has the twofold advantage that all the data are used to evaluate the *H* probabilities and that the values which result minimize the sum of the squared differences between the theoretical and obtained cell frequencies.

Method II may be described concretely by obtaining the solution for a, the probability of the Position Preference *H*. From the 32 equations those containing the variable a are selected. These are

$$\begin{aligned}
a + c + f + r &= Q_1, \\
a + e + r &= Q_2, \\
a + d + r &= Q_3, \\
a + r &= Q_4, \\
a + c + e + p_2 + r &= Q_5, \\
a + f + p_3 + r &= Q_6, \\
a + d + p_3 + r &= Q_7, \\
a + p_2 + r &= Q_8,
\end{aligned}$$

where each Q_a is the relative frequency with which the particular Outcome Sequence (given the Reward Sequence and first-trial outcome) has occurred. In Fig. 1, for example, $Q_1 = 48/(48 + 8 + 25 + 30)$. Adding these eight equations yields

$$8a = \sum_{a=1}^{8} Q_a - 2(c + d + e + f + p_2 + p_3 + 4r).$$

Applying Eq. (1) and solving for a produces

$$a = \frac{\Sigma Q_a}{6} - \frac{1 - b}{3}.$$

The solution for a is thus seen to be a function of the obtained proportions and the variable b. Solving for b in the analogous manner gives

$$b = \frac{\sum_{b=1}^{8} Q_b}{6} - \frac{1 - a}{3},$$

where the Q_b are the obtained proportions in the equations containing b. These last two equations may be solved for a and b. The solutions are

$$a = \frac{3\sum_{a=1}^{8} Q_a + \sum_{b=1}^{8} Q_b}{16} - \frac{1}{2};$$

$$b = \frac{3\sum_{b=1}^{8} Q_b + \sum_{a=1}^{8} Q_a}{16} - \frac{1}{2}.$$

Analogous solutions for each of the other variables are obtained in a similar manner. The resulting equations produce values which minimize the square of the difference between the empirical and theoretical proportions (see Levine, 1959a). A limitation must be noted, however. If some of the Reward Sequences are not permitted to occur in the experiment (e.g., if the correct object is not permitted to remain on the same side for all three trials of a problem) then some of the Q_i are indeterminate and Method II is not applicable.

In summary, it is possible to solve for the H probabilities by first converting the 32 possible sequences to 32 equations in which the H probabilities are the unknowns. The H probabilities may then be obtained by two methods. One may either select nine equations and solve for the nine unknowns (Method I) or one may employ all the data for a least-squares solution (Method II). Method I is useful for testing the model by inserting the obtained values into the other equations; Method II is best employed when the focus is upon the H

functions during the course of an experiment. An empirical problem to be considered is the relationship between the estimates produced by the two methods.

D. Analysis of Object-Discrimination-Learning-Set Data

1. The *H* Functions

Method II was applied to the data of an ODLS experiment with young rhesus monkeys (Harlow *et al.* 1960) which investigated the effects of age upon acquisition (see Chapter 11 by Zimmermann and Torrey in Volume II). Three groups of 10 monkeys each received 600 six-trial problems starting at 60, 90, or 120 days of age. The learning-set functions showing percentage correct on trial 2 for blocks of 100 problems are plotted in Fig. 2. The 60- and 90-day groups were not

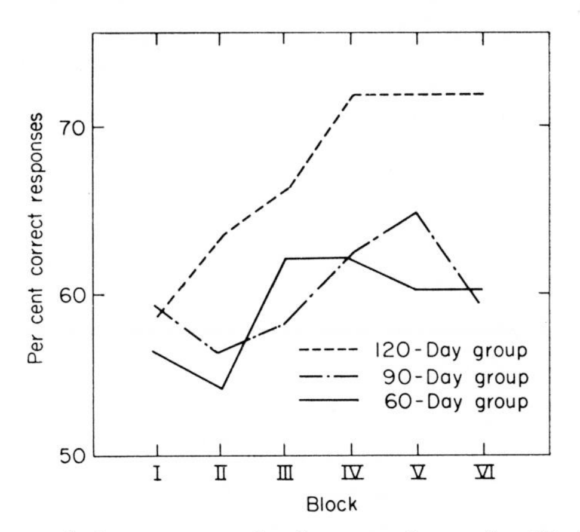

Fig. 2. Performance of three groups in the experiment by Harlow *et al.* (1960) as measured by the percentage of correct responses on trial 2 for each 100-problem block.

significantly different, but the 120-day group was reliably more proficient than either of the others.

The *H* analysis (Method II) was applied to each block for each of the three groups, producing 18 separate analyses. The first finding was that *b*, *d*, *e*, and *f* consistently fluctuated around zero. This was true for each variable in each of the 18 blocks, i.e., 72 measures were taken of variables which appear to have zero strength. The mean of the measures is —0.002, with $\sigma = 0.02$. This result means that these young rhesus monkeys manifested no systematic tendencies toward alternation (H_b and H_d) and no outcome-contingent tendencies directed toward the positions (H_e and H_f). Analysis of data from two other experiments (Levine *et al.*, 1959; Schrier, 1958) indicates that this finding holds for adult rhesus monkeys as well (Levine, 1959b).

The remaining *H*s are plotted in Fig. 3. The p_2 and p_3 values have been added together to give the percentage of problems manifesting a learning *H*, and are plotted in the upper-left-hand quadrant. The differences here reflect the differences seen in Fig. 2, that the 120-day group showed more problem-solution behavior throughout than the 60- and 90-day groups.

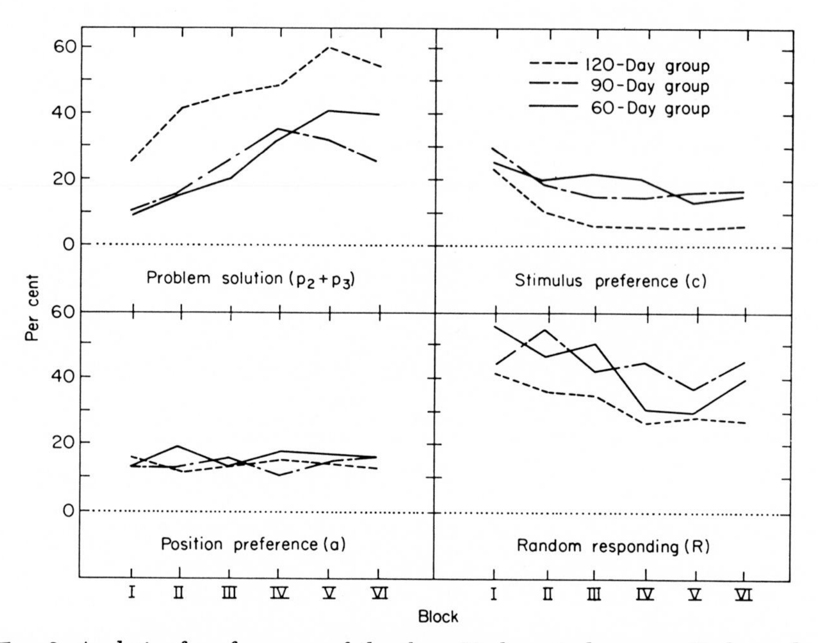

FIG. 3. Analysis of performance of the three Harlow *et al.* groups. Each quadrant shows the percentage of problems in which the behavior was determined by the indicated *H*.

Consideration of the remaining three *H*s suggests some of the sources of these differences. Position Preference is striking in that the *H* has the same strength for the three groups. Thus, the difference among groups in over-all performance is not attributable to H_a. The basis for the difference among the learning-set functions may be seen in the right-hand side of Fig. 3. This difference is attributable specifically to the differences in Stimulus Preference and in the Residual *H*.

In addition to providing more detail about the difference among the groups, Fig. 3 shows the various processes that underlie the learning-set functions. Position Preference is again unique: it does not decrease but has a constant probability of about 0.18. A comparison of this *H* with the other two nonsolution *H*s shows that the *H*s do not extinguish simultaneously or at the same rate.

2. The Proportion of Variance Explained

The application of Method II to describe the *H* structure underlying a learning-set curve carries with it the discomforting problem of validation. One seeks reassurance that numbers such as $a = 0.25$ are not merely artificial results distilled from the mathematical apparatus, but are intimately related to behavior. Some tests are needed to evaluate how well the model describes the data.

It was stated earlier that any theoretical system should account for the frequency patterns encountered when data are organized into matrices as in Fig. 1. Method I may be applied for this purpose. The method was applied to the results of the three groups just discussed. The procedure generally would be to select 9 of the 32 equations corresponding to the 32 sequences, to solve these 9 equations for the 9 *H* probabilities, and to insert these obtained values into the remaining 23 equations, thereby determining how closely the 23 corresponding frequencies can be predicted. In the application to be discussed, however, certain changes were made in this procedure. First, proportions rather than frequencies were predicted. That is, for any sequence (e.g., AAA, $+_1+_2+_3$), instead of predicting the number of occurrences, $n(\mathrm{A}+_1\mathrm{A}+_2\mathrm{A}+_3)$, the proportion of occurrences, $n(\mathrm{A}+_1\mathrm{A}+_2\mathrm{A}+_3)/n(\mathrm{A}+_1\mathrm{AA})$, was predicted. This change was necessary because the Reward Sequences were unequal in the experiment under consideration. Predicting the proportion will control for this variation. If frequencies were predicted here, the fits to the data would be artifactually improved. Second, the stability of the estimates was increased by estimating each *H* probability twice. That is, two sets of simultaneous equations were first selected, permitting two independent evaluations of each *H*. The mean of the two values was taken as the estimate of the *H* probability. Third, eight *H*s instead of nine were considered. This change was introduced because doubling the number of equations for estimation reduces the number of sequences which may be predicted. "Lose-stay-Win-shift with respect to position" was dropped from consideration. This is equivalent to the assumption that $f = 0$. This particular *H* was omitted because it was the "least natural" of the *H*s, and because in all previous work its probability had, in fact, turned out to be close to zero. These changes meant that one half of the data (actually 17 sequences—see below) was to be predicted from the other half of the data.

To avoid any risk of arbitrariness, it was necessary to introduce an automatic procedure for selecting the 16 sequences from which the estimations were to be made. Three criteria were established for this purpose. (1) Wherever possible the sequences were restricted to those ending in an error on the third trial ($-_3$). It was possible to meet this restriction in evaluating all H probabilities other than p_2 and p_3. Twelve sequences ended in $-_3$, and three other sequences, two a function of the variable p_3 and one of the variable p_2, ended in $+_3$. (The sixteenth sequence, for the second evaluation of p_2, was obtained in a different manner, as described below.) In effect, the procedure closely followed was the prediction of the right-hand matrix in Fig. 1 from the left-hand matrix. (2) A second estimate of p_2 was obtained from

$$p_2 = 1 - (a + b + c + d + e + p_3 + R).$$

(3) Those sequences were selected which could be produced only by one, two, or in a few instances, three Hs.

The 15 sequences fulfilling these criteria, as well as the associated equations, are shown in Table I. The data employed are from block I of the 90-day group. This table provides a concrete instance of the manner in which the estimates were obtained. An example of the way the predictions were obtained will be given for sequence $A+_1A+_2A+_3$. It can been seen from Fig. 1 that

$$a + c + e + p_2 + r = P\,(+_2+_3 \mid A+_1AA).$$

TABLE I

THE H ANALYSIS BY METHOD I FOR BLOCK I OF THE 90-DAY GROUP

Sequence	Equations		Solution
A+ A− A−	$r = 0.10$	$r = 0.13$	$r = 0.13$
A+ B+ A−	$r = 0.16$		
A− A− B−	$c + r = 0.41$	$c + r = 0.38$	$c = 0.25$
A− B− B−	$c + r = 0.29$		
A+ A− B−	$b + r = 0.12$	$b + r = 0.14$	$b = 0.01$
A+ B+ B−	$b + r = 0.15$		
A− A+ A−	$d + b + r = 0.17$	$d + r = 0.16$	$d = 0.03$
A− B+ B−	$d + r = 0.16$		
A− B+ A−	$a + d + r = 0.31$	$a + r = 0.24$	$a = 0.11$
A+ B− B−	$a + r = 0.20$		
A+ B− A−	$e + r = 0.09$	$e + r = 0.10$	$e = -0.03$
A− A+ B−	$e + d + r = 0.13$		
A− A− A+	$p_3 + r = 0.13$	$p_3 + r = 0.16$	$p_3 = 0.03$
A− B− A+	$p_3 + r = 0.18$		
A− B+ A+	$p_2 + r = 0.18$	$p_2 = 0.05$	$p_2 = 0.06$
	$p_2 = 1 - (a + \ldots + 4r)$	$p_2 = 0.08$	

The theoretical values for block I of the 90-day group (taken from Table I) are

$$0.11 + 0.25 + (-0.03) + 0.06 + 0.13 = 0.52.$$

This sum is compared to the obtained value given by

$$P(+_2+_3 \mid A+_1AA) = \frac{n(A+_1A+_2A+_3)}{n(A+_1AA)} = \frac{110}{158} = 0.70.$$

The analysis was applied to each block of data for each of the three groups shown in Fig. 2. An example of the results may be seen in Fig. 4,

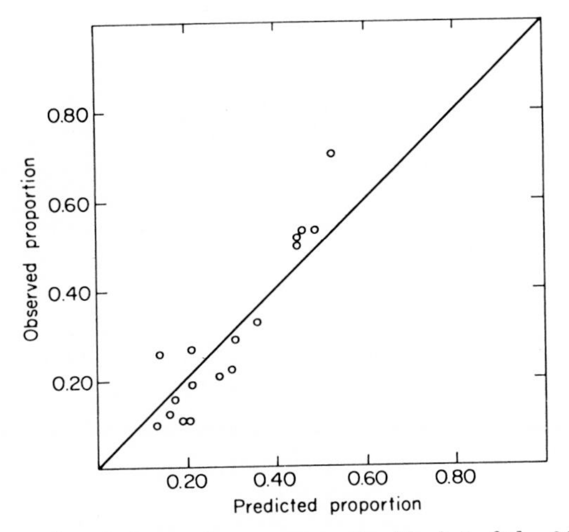

FIG. 4. Predicted and observed proportions for block I of the 90-day group.

which shows the observed and predicted proportions for block I of the 90-day group. Both scales are plotted in the same units, so that perfect prediction would yield 17 points lying along the 45° line. It will be seen that the points distribute themselves in reasonable fashion along this line.

A statistic was devised to quantitatively describe the accuracy of prediction in terms of the proportion of variance explained (P.V.E.) by the predicted values. This is given by:

$$\text{P.V.E.} = 1 - \frac{\sigma_{o \cdot p}^2}{\sigma_o^2},$$

where σ_o^2 is the variance of the observed values, and $\sigma_{o \cdot p}^2$ is the variance around the 45° line, i.e., $\sigma_{o \cdot p}^2 = \sum^{n}(o - p)^2/n$. The symbols o and p

represent the observed and predicted proportions, respectively, and n represents the number of points.[4] For the data in Fig. 4 the P.V.E. = 0.84.

Eighteen P.V.E. were obtained, one for each of the blocks of data in Fig. 2. The mean is 0.85; the range is 0.67 to 0.94. In general, the predictions account well for the observed values. This, of course, means that the model accounts well for the patterns of frequencies observed when the data are organized as in Fig. 1.

3. A Comparison of the Two p_2 Estimates

To determine each of the 18 P.V.E., H_{p_2} was evaluated by two very different formulas (see Table I). One estimate came from a sequence:

$$p_{2_\alpha} = \frac{n(\text{A}-_1\text{B}+_2\text{A}+_3)}{n(\text{A}-_1\text{BA})} - r;$$

the second estimate was obtained by subtraction after all the other H probabilities were obtained:

$$p_{2_\beta} = 1 - (a + b + c + d + e + p_3 + 4r).$$

The relation between the two estimates for each of the 18 blocks of data is shown in Fig. 5, where it can be seen that the points are closely distributed without obvious bias around the line denoting equality. The estimates are close to each other (the mean difference, $|p_{2_\alpha} - p_{2_\beta}|$, is

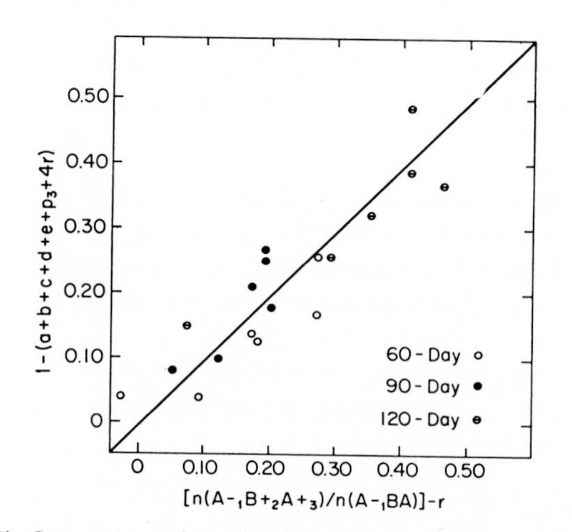

Fig. 5. Comparison of each of the 18 pairs of estimates of p_2.

[4] The statistic P.V.E. is analogous to the square of the correlation coefficient, r^2. P.V.E. $\leq r^2$.

0.05) despite the fact that p_{2_β} would be in error to the extent that the sum of the other estimated H probabilities was in error.

4. A COMPARISON OF METHOD I AND METHOD II

The H probabilities for each of 18 blocks of data from the study by Harlow *et al.* were obtained twice, first by a least-squares method (Method II) then by the solution of systems of simultaneous equations (Method I). For each block the correlation was calculated between the two sets of estimates of the H probabilities. The lowest correlation coefficient was 0.96. The mean was 0.98. Thus, although the two methods involve different algebraic procedures, the H probabilities that they yield are practically identical.

III. AN EXPERIMENT DESIGNED TO TEST THE MODEL

A. Introduction

Heretofore, the tests used data from a conventional ODLS experiment which had been executed without prior considerations of relevance to the model. In the present section a new type of ODLS experiment will be described, one which is suggested by the analysis of behavior into Hs and which permits testing of predictions generated by the model.

Each problem of the ODLS situation may be characterized as a two-person game (Luce & Raiffa, 1957) in which the two opponents are the subject and the experimenter. Each problem corresponds to one play of the game, the subject's strategies are the Hs, and the outcomes are the transfer of 0 or 1 raisin from the experimenter to the subject on each trial. The probabilities of the Hs are, of course, obtained from the analysis described thus far. The previous sections have shown that five Hs predominate: H_a, H_c, H_{p_2}, H_{p_3}, and H_R. The experimenter, on the other hand, may be considered as having a single strategy (Sy)[5]: prearranged object baiting. That is, on any problem the experimenter decides in advance, independent of the subject's responses, which of the two objects will be associated with the raisin, and maintains this association for every trial of the problem. It is possible for the experimenter to employ a second Sy a certain proportion of the time, such that the new Sy interacts with the various Hs differently from the conventional Sy and such that no cues are available to the subject to indicate which Sy is operating. The new Sy in this experiment has this procedure: the experimenter baits both objects on trial 1 and continues on subsequent trials to bait the object selected by the subject on trial 1.

[5] The symbol "Sy," except that it refers to the experimenter's alternatives, will have the same formal status as "H." The distinction is made only to reduce confusion.

Table II shows the situation in which the experimenter employs the conventional *Sy* two-thirds of the time and the new *Sy* one-third of the time, and in which the subject employs the five *H*s mentioned above. The numbers in each cell show, for trial 2, the average value in raisins to the subject of the various *H* and *Sy* combinations. The matrix for trial 3 is the same except that the numbers in both H_{p_3} cells are 1.0 instead of 0.0.

If the probabilities of the *H*s were known in advance, then the percentage of correct responses made by the subject for each of the experimenter's two *Sy*s could be estimated from expected-value functions. For

TABLE II

EXPECTED VALUE IN RAISINS ON TRIAL 2 FOR EACH *H-Sy* COMBINATION

H		Conventional *Sy* (2/3)	New *Sy* (1/3)
Position Preference	(a)	0.5	0.5
Stimulus Preference	(c)	0.5	1.0
Win-stay, etc.	(p_2)	1.0	1.0
Third-trial Learning	(p_3)	0.0	0.0
Residual	(R)	0.5	0.5

NOTE: Probabilities of the various *H*s and *Sy*s are shown in parentheses.

example, under the conventional *Sy* the proportion of raisins transferred (or, equivalently, the proportion of correct responses) on trial 2 is obtained as follows. Whenever H_{p_2} occurs, the response on trial 2 is correct; whenever H_{p_3} occurs, the response on trial 2 is never correct; and whenever one of the other *H*s occurs, the response on trial 2 is correct half the time (cf. the conventional-*Sy* column in Table II). Percentage correct on trial 2, under the conventional *Sy*, then, would be estimated by

$$(4) \qquad (\% +_2)_{\text{con.}} = 100\left[p_2 + \frac{1-(p_2+p_3)}{2}\right].$$

Similarly, for trial 3,

$$(5) \qquad (\% +_3)_{\text{con.}} = 100\left[p_2 + p_3 + \frac{1-(p_2+p_3)}{2}\right].$$

Under the new *Sy*, the subject is also rewarded whenever it holds H_c. Therefore, percentages correct on trials 2 and 3 are

$$(6) \qquad (\% +_2)_{\text{new}} = 100\left[p_2 + c + \frac{1-(p_2+p_3+c)}{2}\right], \text{ and}$$

$$(7) \qquad (\% +_3)_{\text{new}} = 100\left[p_2 + p_3 + c + \frac{1-(p_2+p_3+c)}{2}\right].$$

Of course, the *H* probabilities are never given *a priori* but must be obtained from the data under consideration. The advantage of the two-*Sy* design described here is that the *H*s can be evaluated from the data obtained under the conventional *Sy*, and the percentage correct can then be predicted for performance under the new *Sy*. That is, percentage correct on trials 2 and 3 under the new *Sy* is predicted by Eqs. (6) and (7), respectively, in which p_2, p_3, and c are evaluated from performance under the conventional *Sy*.

B. Method

Sixteen pig-tailed macaques (*Macaca nemestrina*) were used in this study. From their weight and size it appeared that these monkeys were all over two years of age. They had had no previous experience in any discrimination experiments. Discrimination training was conducted in the Wisconsin General Test Apparatus (see Chapter 1 by Meyer *et al.*). The stimuli to be discriminated were pairs of three-dimensional objects randomly selected from a set of 2,000 objects.

The experiment consisted of 576 problems each lasting for three trials. Two-thirds of these problems were presented under the conventional *Sy*; the remaining third of the problems were presented under the new *Sy*. The *Sy*s were randomly assigned to the problems within a framework of restrictions designed to maintain the two-to-one ratio over sets of twelve problems and to assure that all eight permutations of position sequence occurred in the eight conventional problems of each set.

Under the conventional *Sy* all trials followed the standard procedure. The object designated as correct was placed over the baited foodwell, the other object over the empty well, and the tray presented to the monkey. After the monkey displaced one of the objects, and, if the response was to the correct object, took the reward, the tray was withdrawn. This noncorrection technique was used throughout.

Under the new *Sy* both objects were baited with a raisin on the first trial. Only the object selected was then baited on trials 2 and 3. In all other details the standard procedure was followed.

C. Results and Discussion

Table III shows the probabilities of the *H*s as evaluated from four successive 96-problem blocks of conventional problems. There are two noteworthy differences between the trends here and the curves shown for the young rhesus monkeys in Fig. 3. Here Position Preference virtually disappears after the first block of problems, and Stimulus Preference is high for the first two blocks of problems and drops abruptly

TABLE III
H PROBABILITIES OBTAINED FROM SUCCESSIVE BLOCKS OF 96 CONVENTIONAL PROBLEMS

H		Block 1	2	3	4
Position Preference	(a)	0.13	0.03	0.01	0.02
Stimulus Preference	(c)	0.13	0.09	0.01	0.01
Win-stay, etc.	(p_2)	0.17	0.28	0.46	0.55
Third-trial Learning	(p_3)	0.06	0.08	0.10	0.07
Residual	(R)	0.51	0.52	0.41	0.35

for the last two. Whether the change in results reflects the different ages or species of the animals, or the difference in experimental designs, cannot, at present, be determined.

The behavior of Stimulus Preference is reflected in Fig. 6, which shows the learning-set functions obtained under the conventional *Sy* and under the new *Sy*, and the predicted function for the new *Sy*. Theo-

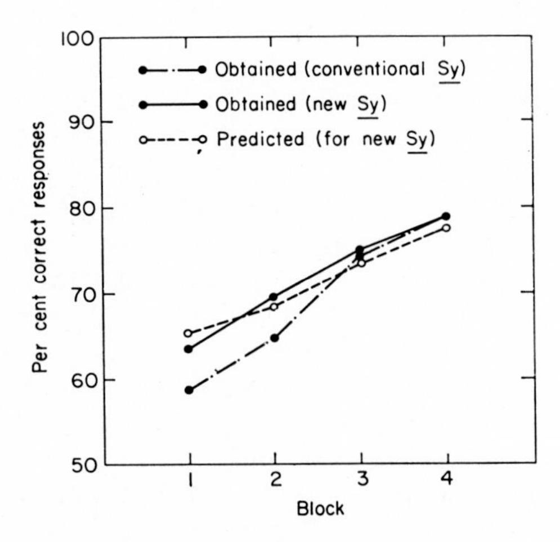

FIG. 6. Obtained percentage of correct responses under the conventional and new *Sys*, and predicted percentage for the new *Sy* (for trials 2 and 3 combined).

retically the difference between the two empirical functions should equal $100(c/2)$. [Subtracting Eq. (4) from Eq. (6) will prove this theorem for trial 2; subtracting Eq. (5) from Eq. (7) will prove it for trial 3.] The obtained differences for the four blocks of problems were 4.9, 4.9, 0.4, and 0.0; the corresponding values of $100(c/2)$ are 6.5, 4.5, 0.5, and 0.5.

The similarity between these two sets of numbers accounts for the close fit of the performance under the new *Sy* by the predicted performance. This correspondence is further analyzed in Fig. 7, which shows the predicted and obtained functions for trials 2 and 3 separately.

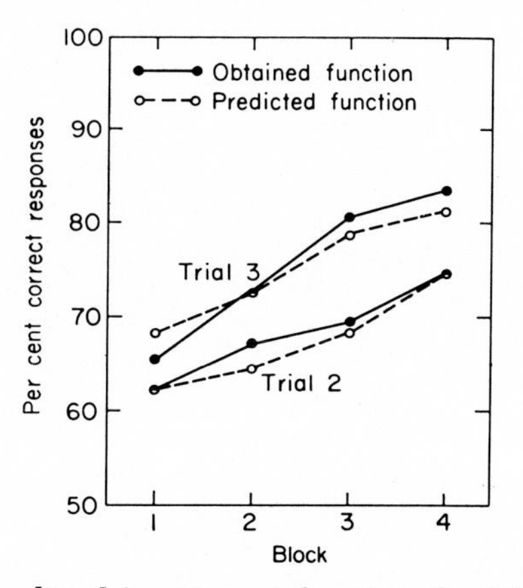

FIG. 7. Obtained and predicted learning-set functions for trials 2 and 3 separately.

A test of goodness of fit suggested by Lindquist (1953, pp. 344–346) was applied to these data. The predicted and obtained results were not significantly different.

Another investigation was made of the effects of differences in c upon the quality of predictions. The value of c was computed for each monkey during problems under the conventional *Sy*. The monkeys were then rank-ordered according to their c-values, and every four monkeys were assigned to a different group. The mean c-values for the four groups were: −0.05, 0.07, 0.17, and 0.26. The *H* analysis was applied, for each group separately, to the conventional-*Sy* data. The predicted percentage correct under the new *Sy* could then be computed from Eqs. (6) and (7), and compared to the obtained percentage correct. The results, for trials 2 and 3 combined, are shown in Fig. 8. Again, the close fit is striking. By t test, the largest difference, that for the group with $c = 0.26$, was not significant.

The general conclusion from this experiment is that a hypothesis model can account for the difference in performance under the two *Sys*. More specific conclusions are that the difference in performance is approximated by $100(c/2)$, and that the expected-value function, using the calculated H values, gives a good estimate of percentage correct.

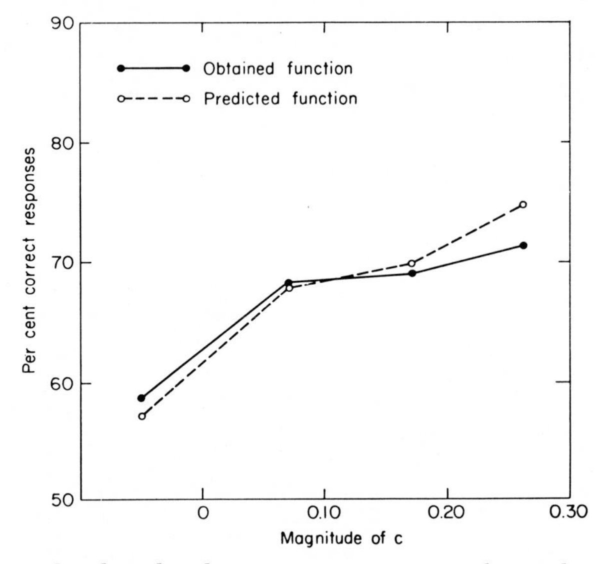

FIG. 8. Obtained and predicted percentage correct on trials 2 and 3 combined for the four groups differing in magnitude of Stimulus Preference (c).

IV. ANALYSIS OF INTRAPROBLEM ODDITY DATA

A. Introduction

The next application of the model will be to data on oddity learning presented previously (Levine & Harlow, 1959). In this learning situation a row of three objects was presented on each trial, two of the objects being identical, the third different, or *odd*. The odd object always covered food. A series of 12-trial problems was presented to a group of rhesus monkeys. A problem, as in the ODLS experiment, was defined as a set of trials with the same objects, and a new problem meant that two new identical pairs of objects were randomly selected from a heterogeneous collection. On each trial, the odd object was randomly chosen from the four objects (two identical pairs) of the current problem. The position of the odd object was also randomized, with an added restriction that the odd object be always on the right or left, never in the center. Response to the center object was prevented by having a transparent screen between the object and the monkey. The oddity problem was thus converted to a two-response situation (see Chapter 5 by French).

The chief findings from this experiment were: (a) Learning was not manifested within a problem. Figure 9 shows the within-problem "learning curve" averaged over the first block of 36 problems. The percentage of correct responses clearly fluctuated around a constant value from trial to trial.

(b) Learning was manifested over blocks of problems, but the improvement was very small. After 3,000 trials the group made a mean of 62% correct responses. Another change over blocks, however, was that the within-problem fluctuations tended to disappear.

A finding by Moon and Harlow (1955) suggested an explanation for the large fluctuations in the first block of problems. These authors had shown that "Win-stay-Lose-shift with respect to the object" predominates in the oddity-learning-set experiment when that experiment follows ODLS. Levine and Harlow, lacking a model for quantitative *H* analysis, simply asked what would happen if all monkeys always showed this

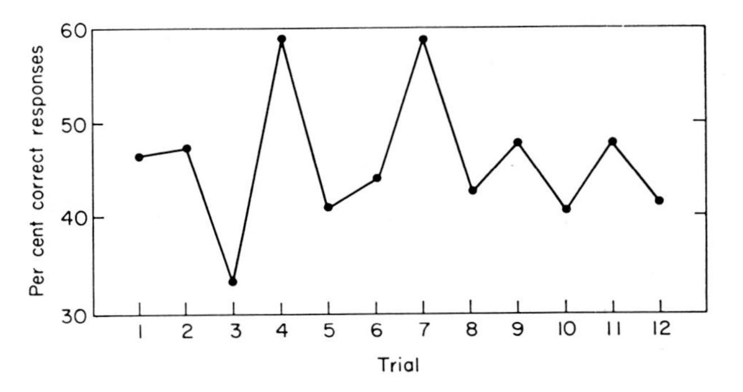

FIG. 9. Percentage of correct responses on each trial during the first 36 oddity problems.

tendency. They generated the theoretical sequence of responses which must appear given this *H*, and determined the percentage of time that the odd object would be chosen on each of the 12 trials. This hypothetical within-problem learning curve compared favorably with the curve obtained during the initial block of problems, showing a pattern of ups and downs similar to that in Fig. 9.

Useful as this demonstration was, two problems remained. The first was that the value of the *H* strength was not derived from any systematic analysis of response patterns. It was simply assumed that the probability of the *H* was 1.0, and that, in effect, no other *H*s existed. This arbitrary assumption led to an exaggeration of some of the predicted fluctuations.

The second problem was that the change in the shape of the curve over blocks of problems could not be treated. Even the additional assumption of a gradual decrease in the probability of the postulated *H* would not account for some of the changes.

A more satisfactory approach would be to determine first the probability distribution over the total set of possible *H*s and to derive from this the within-problem curves. The earlier sections of this chapter have suggested that such an analysis is feasible, and the application will be made in this section.

B. Experimental Method

Seven female rhesus monkeys served as the subjects. They had previously reached near-asymptotic performance in the course of approximately 5,000 trials of an ODLS experiment.

Pairs of identical multidimensional objects, selected from a collection of 1,000 different pairs, were randomly grouped into sets of two, each such set comprising the stimulus objects for one 12-trial problem. The object designated as odd was placed on the right or left of the test tray on any trial according to a sequence adapted from Gellermann (1933b). The sequence lasted for 18 problems (216 trials) and was repeated every 18 problems. The object to be odd on any one trial was stipulated by labeling each pair of the set as *X* or *Y* and then developing a random *X-Y* sequence. This sequence also contained the Gellermann restrictions and also lasted for 18 problems. There was, then, a cycle of position and object sequences which repeated every 18 problems. A series of 252 problems was presented.

C. The Analysis

The first task was to adapt to 12-trial problems the method of analysis described above for 3-trial problems. The most straightforward solution was taken. A 12-trial problem was divided into four sets of 3 trials and each such set was treated as the unit over which systematic behaviors were to be determined. That is, assumption (3), that a single *H* lasted through the first 3 trials of a problem, now was applied to a 3-trial unit within a problem.

This simple modification may be unrealistic. It is equivalent to assuming that the monkey changes an *H* only after trials 3, 6, or 9, and not after any of the other trials. The best alternative to this assumption, however, is a reconstruction of the model for intraproblem analysis. Such reconstruction remains a project for the future.

The analysis was applied to blocks of 36 problems, or 144 units per monkey. The *H*s selected for the analysis were those which had proved to be important in the preceding treatments. Of course, another *H*, manifested as systematic response to the odd object, had to be added. Thus, H probabilities were evaluated for the following:

H_a : Position Preference,
H_c : Stimulus Preference,
H_p : Win-stay-Lose-shift with respect to the object,
H_o : Preference for the odd object,
H_R : Residual.

Because the ultimate aim, constructing a theoretical within-problem learning curve, involves the interaction between the Hs and the left-right and X-Y sequence followed by the odd object, it was necessary to have a more refined analysis than previously. For example, it was necessary to determine not only the amount of systematic response to position, i.e., $P(H_a)$, but to the right position as distinct from the left position. If the Gellermann sequences had been counterbalanced so that for each of the 12 trials the odd object was on the right side exactly 50% of the time, then such a refinement would not be necessary. However, such detailed counterbalancing had not been accomplished. Over the 18-problem positional sequence, the odd object was on the right side 10 times on trial 1, 8 times on trial 2, 12 times on trial 3, etc. If the monkeys had a bias toward the right side they would make more than 50% correct responses on trials 1 and 3 when they employed H_a. In fact, Position Preference directed to the right side ($H_{a \cdot r}$) would produce 67% correct responses on trial 3; a preference for the left side ($H_{a \cdot l}$) would produce 33% correct responses on trial 3. Therefore, $H_{a \cdot r}$ and $H_{a \cdot l}$ were separately evaluated, with

$$P(H_a) = P(H_{a \cdot r}) + P(H_{a \cdot l}).$$

The same considerations apply to H_c since objects called X (or Y) by the experimenter might be preponderantly preferred. The number of times that the odd object was X on each trial would then be important in determining the percentage correct on each of the 12 trials of the intraproblem curve. Therefore, preferences for X objects ($H_{c \cdot x}$) and Y objects ($H_{c \cdot y}$) were separately analyzed, with

$$P(H_c) = P(H_{c \cdot x}) + P(H_{c \cdot y}).$$

To observe the changes in H strengths and in the intraproblem learning curves throughout the experiment, the first (F: problems 1–36), middle (M: problems 109–144), and last (L: problems 217–252) blocks of 36 problems were subjected to analysis. Method I alone could be applied. Certain L-R and X-Y sequences never occurred, making the application of Method II impossible.

D. Results of the Analysis

The *H* probabilities are presented in Fig. 10. Several features of this figure deserve attention. Position Preference and Stimulus Preference had almost the same magnitude found in the earlier analyses of ODLS data. The striking feature is that these monkeys had already achieved near-perfect performance on ODLS, which implies that these two *H*s

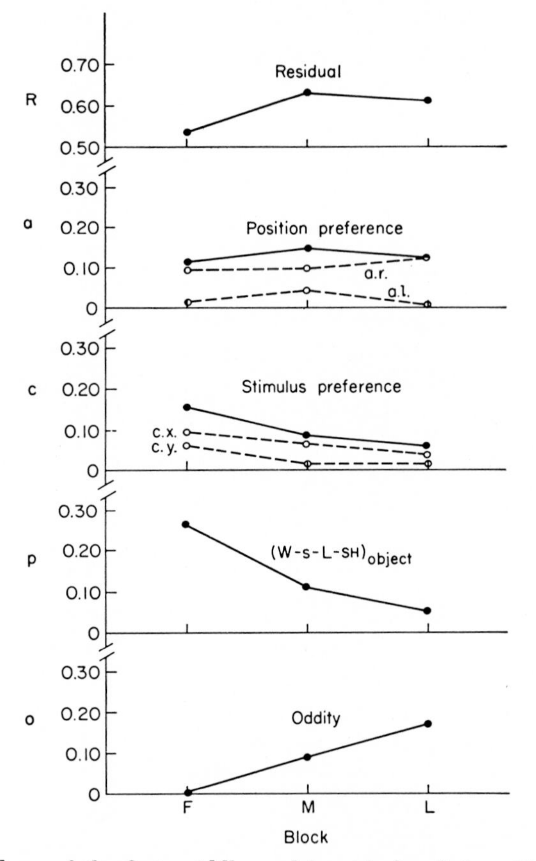

FIG. 10. *H* analyses of the first, middle, and last blocks of the oddity experiment.

must have extinguished in the course of that earlier experiment. They reappeared in this subsequent experiment. Restle (1958) has suggested that such an effect, reappearance of extinguished behaviors when the problem is changed, might occur (see Chapter 2 by Miles).

Position Preference showed the peculiarity that it did not decrease over the course of the experiment. The same finding, it will be recalled, appeared in the analysis of ODLS for the three groups of infant rhesus monkeys, but not for the adult pig-tailed monkeys. The ODLS was a difficult task for the infants, as was the oddity task for the present subjects. For adult monkeys, however, ODLS is relatively easy. Rate of extinction of Position Preference may be a sensitive indicator of task difficulty.

The refined analysis of Position Preference revealed that practically all the preference manifested was for the right side. This will become important when the intraproblem curves are considered. A similar effect, though less extreme, appeared for Stimulus Preference.

Finally, a result ominous for the aim of theoretical consistency appears: $\Sigma P(H_i) > 1.0$. For the three blocks the sums were 1.08, 1.07, and 1.04, respectively. The source of this effect and corollary problems are discussed below.

E. The Intraproblem Curve

Using the data from the H analysis, a theoretical curve showing percentage of correct responses on each of the 12 trials was constructed. Exactly the same L-R and X-Y sequence was presented to each animal twice during each block of 36 problems. From these sequences, the number of times in the block that the odd object was on the right (or left) on trial 1, on trial 2, etc., was determined. This information gives the number of times (out of 36) that $H_{a \cdot r}$ (or $H_{a \cdot l}$) was rewarded on trial 1, trial 2, etc. The number of times that $H_{c \cdot x}$ (or $H_{c \cdot y}$) was rewarded on any trial was similarly determined by counting the number of times that the X (or Y) object was odd for that trial.

The number of times H_p was rewarded on each trial was determined by following this rule: If the object odd on trial n is also odd on $n + 1$, the response on $n + 1$ will be rewarded; otherwise not. This rule gives the outcome for any trial of any problem except the first trial. For the first trial it was assumed that half the responses were to the odd object; for the other eleven trials, the theoretical number of rewarded responses, given H_p, was counted.

The treatment of H_o was simple: so long as this H is held every response is rewarded. Given this H there will be 36 rewards on trial 1, on trial 2, etc., over the block of problems.

For H_R it was assumed that exactly one-half of the responses are to the odd object on each trial.

Each number so obtained was divided by 36 to yield, for the jth trial, the probability of reward given the H. This will be symbolized as $P(+_j|H_i)$. The ultimate aim was a curve showing $P(+_j)$, the over-all (unconditional) probability of reward on the jth trial. This is given by

$$(8) \qquad P(+_j) = P(H_{a \cdot r})\, P(+_j \mid H_{a \cdot r}) + P(H_{a \cdot l})\, P(+_j \mid H_{a \cdot l}) + \ldots + P(H_R)\, P(+_j \mid H_R).$$

The left-hand probability in each term is, of course, given by $a \cdot r$, $a \cdot l$, . . . , R, respectively. These are the H probabilities found previously and shown in Fig. 10. The right-hand probability was obtained for each H in the manner just described.

Intraproblem curves were constructed, for each 36-problem block, by solving Eq. (8) for each trial. The results were dismaying. In general, the within-problem curves were too flat. While the pattern for the first block was generally reproduced, the peaks and valleys were systematically less extreme than those obtained. Also, the theoretical curve was too high: the obtained curve fluctuates around a mean of 0.46; the theoretical curve around 0.51.

These defects in the predictions could occur if a single parameter were misevaluated. If r, as estimated from the frequency of occurrence of those sequences not produced by any of the systematic Hs, were too large, then: (a) The error in R ($= 8r$ in the oddity model) would be amplified and would produce a flattened curve (since R is a "weighting" for flatness–$P(+_j)$ is 0.5 at *every* trial given this H). (b) The effect seen earlier, $\Sigma P(H_i) > 1.0$, would be produced. (c) As a result, the curve would be too high. Also, (d) the probability of each of the systematic Hs, estimated by subtracting the calculated value of r from $Q_i = P(H_i) + r$, would be too low.

Accordingly, the procedure was revised to determine r from the Q_i, incorporating the assumption that $\Sigma P(H_i) = 1.0$. The general rationale was as follows: For any of the systematic Hs, an equation of the form $Q_i = P(H_i) + r$ can be obtained by averaging data from selected cells. If there are n systematic Hs there are n resulting equations, but with $n + 1$ unknowns (since r appears in each equation). An additional equation follows from Assumptions (2) and (5):

$$a \cdot r + a \cdot l + \ldots + R = 1.0.$$

Thus all the probabilities may be obtained.

This procedure, applied to the present data, yielded a reduced value of R and a corresponding increase in the other H probabilities of about 5%. Substituting the H probabilities derived by this modification into Eq. (8) to determine $P(+_j)$ yielded the results shown in Fig. 11. It is evident that the correspondence between the theoretical and obtained curves is good. The flattening across blocks is predicted, as are the locus and magnitude of the fluctuations. One detail, in particular, reveals the virtues of an H model for explaining the obtained curves. During the first block the lowest percentage of correct responses occurred at trial 3. During the last block the minimum had shifted to trial 5. Prediction of this shift by the theoretical curves is explained by consideration of the Hs predominating at each block. During the first block, H_p was the maximum systematic H. The probability of a reward given this H is lowest at trial 3: $P(+_3|H_p) = 0.11$. As H_p declined over problem blocks, the depression of trial 3 performance diminished. By the last block, the most important systematic error-producing H was $H_{a.r}$. The probability of a reward given this H is lowest at trial 5: $P(+_5|H_{a.r}) = 0.22$. Therefore, a relative depression was produced at trial 5.

The synthesized curves, then, provide good matches to the obtained results. As good as the comparisons may be, however, the way they were attained was far from an ideal procedure. In one sense the model has been checked: The set of assumptions and the general logic led to

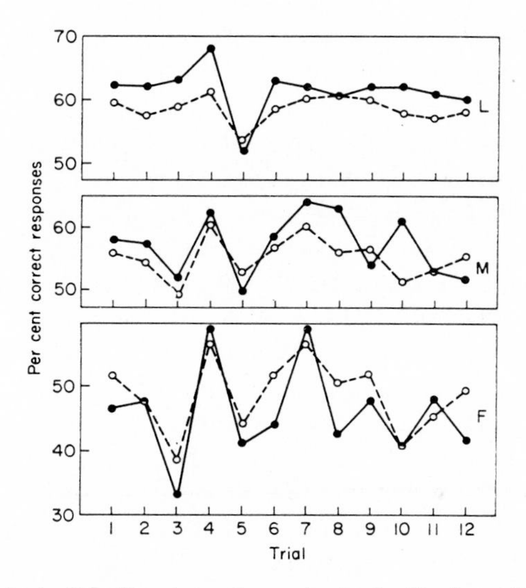

FIG. 11. Obtained (solid lines) and synthesized (broken lines) intraproblem curves for the first, middle, and last blocks of the oddity experiment.

Method I which, for the oddity data, produced unsatisfactory results. Only after seeing these results, diagnosing the problem as a defect in the evaluation of r, and correcting this evaluation were the theoretical curves of Fig. 11 produced. Two facts, however, mitigate the damage to the general ideas of the model. First, the model was elaborated for short problems, application to which yields more satisfactory results (Sections II and III; see also Levine, 1963). Twelve-trial problems were treated by redefining the behavioral unit. The artificiality of this new definition, it may well be asserted, was responsible for the defective predictions obtained. Second, the change in the method of evaluating r was standard and applied uniformly in the production of all the predicted points. Even though the results follow an after-the-fact adjustment, therefore, it is not rash to argue that the bizarre intraproblem curves may be accounted for by the ideas of *H* theory.

V. DISCUSSION

Apart from the detailed results presented in this chapter, the model of *H* behavior has certain general characteristics worthy of note.

(1) The *H* is defined as a mediating process. Intuitively, it may be thought of as a "tendency" of the organism. It is assumed that this tendency is not the systematic behavior itself but is manifested in such behavior. This definition is a change from previous usage which defined the *H* as a behavior pattern (Behar, 1961; Krechevsky, 1932; Levine, 1959b). The shift in definition is not arbitrary but is required by the fact that a single *H* is rarely synonymous with a response pattern. Suppose, for example, that a problem of type AAA (see Fig. 1) is presented, and that the subject responds III, receiving three rewards. One cannot tell from the behavior whether this is an instance of H_a, H_c, or H_{p_2} (among others). If the *H* were defined as the observed pattern of three responses, therefore, the statement that the *H*s are mutually exclusive could not be made. To assume that one and only one *H* has occurred is to assume that some part, some subset, of the environment is controlling the subject's behavior, but that the behavior itself does not reveal the subset. In this sense, the *H* is like the concept of "attention," implying as it does a process which relates behavior to a subset of the environment. When aspects of the environment are confounded, as position, object, and reward are in the AAA problem, one cannot tell just which process is operative, i.e., just which aspect of the environment has been related to behavior.

(2) The *H*, rather than the specific choice response on any one trial, is regarded as the dependent variable affected by the rewards. This point of view has a few advantages. First, the learning-set effect can be treated within the context of a conditioning theory. In the typical ODLS experiment, H_{p_2} is the only *H* that receives 100% reward after trial 1. Levine (1959b) proposed that this feature strengthens H_{p_2} relative to the other *H*s, and, consequently, produces the improvement in interproblem performance. Schusterman (1962) has provided an important confirmation of this hypothesis. He trained chimpanzees on three repeated-reversal problems (see Chapter 5 by French and Chapter 7 by Warren). With this type of problem H_{p_2} receives the maximum rewards. If repeated reversal strengthens H_{p_2} and if a strong H_{p_2} is the basis of successful learning-set performance, then these subjects should show positive transfer when trained on ODLS. Schusterman found that the chimpanzees were at asymptote (almost 90% correct responding on trial 2) in the very first block of ODLS problems. Chimpanzees that had

been trained on object alternation without any sign of learning showed virtually chance responding on block 1 followed by the more typical learning-set function. One may conclude that strengthening H_{p_2} is a sufficient condition for producing efficient problem-solving in the ODLS experiment (see Chapter 2 by Miles).

The second advantage of treating the *H* as the dependent variable is that the paradox of alternation learning (see Chapter 5 by French) is eliminated. Behar (1961) has demonstrated that rhesus monkeys can develop an alternation-learning set. Consider a monkey, near asymptote in such an experiment, that has just responded to the correct object on trial 1 and has received food. This outcome is almost universally regarded as reinforcement of the response. But does it have this function? The monkey typically chooses the other object on the next trial. Therefore, the experimenter has reinforced one response yet has increased the probability that the monkey will make the other response—a puzzling result by most definitions of "reinforcement." The resolution of this paradox is that the reinforcements affect not the choice response but the *H* selected. In this case a series of trials reinforces "Win-shift-Lose-stay with respect to the object."

The third, and perhaps most important, feature of redefining the response is that one is no longer limited to curves of "correct responses." Rather, one may follow and compare the variety of processes that exist, studying not only the "response" (i.e., the *H*) strengthened but the extinction of several others as well.

(3) The *H* analysis generated by the model adds a tool for the developmental and comparative study of behavior and for physiological analysis. An instance of its potential in developmental analysis is shown in Fig. 3, which depicts *H* curves for monkeys of different ages. Hypothesis analyses have been used in the study of development of the rhesus monkey (Harlow *et al.*, 1960) and of the human (Hill, 1961). In comparative research, Levine (1963) showed that Stimulus and Position Preferences, so ubiquitous in studies of nonhuman primates, were absent in adult-human data. For a more direct approach to comparative analysis, see Schusterman (1963). In physiological research, *H* patterns have been correlated with changes in brain chemistry (Rosenzweig *et al.*, 1960) and with lesion locus (Brush *et al.*, 1961).

(4) Along with these results and potentials are several problems that remain to be solved before the model can serve as a comprehensive theory of discrimination learning. First, and most pressing, is the revision needed for intraproblem analysis. The single long discrimination problem has not yet been formally treated. A first attempt, with the 12-trial oddity problem, yielded results of equivocal merit, requiring *a posteriori* adjustment in the measurement process (Section IV). Second is the

need to synthesize conditioning concepts with the ideas of an *H* model. How do *Hs* occur which initially have zero strength (as, for example, in the oddity experiment)? What sort of process brings them into the set of *Hs* from which the subject samples? Also, how does the outcome of each trial interact with the *Hs*? Solution of this latter question should enable one to generate theoretical *H* functions in something like the manner that contemporary mathematical models generate correct-choice-response functions. A start on this question has been made by Restle (1962). Third is the treatment of the Residual *H*. The current treatment, that this *H* is manifested directly in patterns not produced by any other *Hs*, keeps the mathematics unencumbered. The chief defect with this approach, however, is that it leaves out of account momentary sources of error, "slips" by a subject that has the correct (or any other) *H*. Because of this, the Residual *H* as now measured would increase artifactually with longer problems. It is anticipated that a more satisfactory treatment of this *H* will be distilled as data accumulate.

Harlow has urged that "Until the nature of error factors can be experimentally demonstrated, there is little hope for the formulation of an adequate discrimination learning theory" (Harlow, 1950, p. 26). It is clear that the point has not yet been reached for evaluating this thesis. Hypothesis theory must first negotiate severe obstacles. This chapter describes preliminary encounters on the path.

References

Behar, I. (1961). Analysis of object-alternation learning in rhesus monkeys. *J. comp. physiol. Psychol.* **54**, 539.

Brush, E. S., Mishkin, M., & Rosvold, H. E. (1961). Effects of object preferences and aversions on discrimination learning in monkeys with frontal lesions. *J. comp. physiol. Psychol.* **54**, 319.

Gellermann, L. W. (1933a). Form discrimination in chimpanzees and two-year-old children: II. Form versus background. *J. genet. Psychol.* **42**, 28.

Gellermann, L. W. (1933b). Chance orders of alternating stimuli in visual discrimination experiments. *J. genet. Psychol.* **42**, 207.

Hamilton, G. V. (1911). A study of trial and error reactions in mammals. *J. anim. Behav.* **1**, 33.

Harlow, H. F. (1949). The formation of learning sets. *Psychol. Rev.* **56**, 51.

Harlow, H. F. (1950). Analysis of discrimination learning by monkeys. *J. exp. Psychol.* **40**, 26.

Harlow, H. F. (1959). Learning set and error factor theory. *In* "Psychology: A Study of a Science" (S. Koch, ed.), Vol. 2, pp. 492–537. McGraw-Hill, New York.

Harlow, H. F., Harlow, Margaret K., Rueping, R. R., & Mason, W. A. (1960). Performance of infant rhesus monkeys on discrimination learning, delayed response, and discrimination learning set. *J. comp. physiol. Psychol.* **53**, 113.

Hill, S. D. (1961). The chronological age levels at which children solve three problems varying in complexity. Paper read at Psychonomic Society, New York.

Krechevsky, I. (1932). "Hypotheses" in rats. *Psychol. Rev.* **39**, 516.

Leary, R. W. (1958). Analysis of serial discrimination learning by monkeys. *J. comp. physiol. Psychol.* **51**, 82.

Levine, M. (1959a). A model of hypothesis behavior in discrimination learning set. Doctoral dissertation, University of Wisconsin. University Microfilms, Ann Arbor, Michigan, No. 59–1397.

Levine, M. (1959b). A model of hypothesis behavior in discrimination learning set. *Psychol. Rev.* **66**, 353.

Levine, M. (1963). Mediating processes in humans at the outset of discrimination learning. *Psychol. Rev.* **70**, 254.

Levine, M., & Harlow, H. F. (1959). Learning-sets with one- and twelve-trial oddity-problems. *Amer. J. Psychol.* **72**, 253.

Levine, M., Levinson, Billey, & Harlow, H. F. (1959). Trials per problem as a variable in the acquisition of discrimination learning set. *J. comp. physiol. Psychol.* **52**, 396.

Lindquist, E. F. (1953). "Design and Analysis of Experiments in Psychology and Education." Houghton Mifflin, Boston, Massachusetts.

Luce, R. D., & Raiffa, H. (1957). "Games and Decisions." Wiley, New York.

Moon, L. E., & Harlow, H. F. (1955). Analysis of oddity learning by rhesus monkeys. *J. comp. physiol. Psychol.* **48**, 188.

Pubols, B. H. (1962). An application of Levine's model for hypothesis behavior to serial reversal learning. *Psychol. Rev.* **69**, 241.

Restle, F. (1958). Toward a quantitative description of learning set data. *Psychol. Rev.* **65**, 77.

Restle, F. (1962). The selection of strategies in cue learning. *Psychol. Rev.* **69**, 329.

Rosenzweig, M. R., Krech, D., & Bennett, E. L. (1960). A search for relations between brain chemistry and behavior. *Psychol. Bull.* **57**, 476.

Schrier, A. M. (1958). Comparison of two methods of investigating the effect of amount of reward on performance. *J. comp. physiol. Psychol.* **51**, 725.

Schrier, A. M., & Harlow, H. F. (1956). Effect of amount of incentive on discrimination learning by monkeys. *J. comp. physiol. Psychol.* **49**, 117.

Schusterman, R. J. (1962). Transfer effects of successive discrimination-reversal training in chimpanzees. *Science* **137**, 422.

Schusterman, R. J. (1963). The use of strategies in two-choice behavior of children and chimpanzees. *J. comp. physiol. Psychol.* **56**, 96.

Spence, K. W. (1937). Analysis of the formation of visual discrimination habits in chimpanzee. *J. comp. Psychol.* **23**, 77.

Yerkes, R. M. (1916). Ideational behavior of monkeys and apes. *Proc. nat. Acad. Sci.* **2**, 639.

Chapter 4

The Delayed-Response Problem

Harold J. Fletcher

Department of Psychology, University of Wisconsin, Madison, Wisconsin

I. INTRODUCTION

With no close rival, past or present, the delayed-response problem remains the one behavioral test most sensitive to the widest range of experimental treatments. This powerful behavioral assay reveals phylogenetic, ontogenetic, and sex differences; it detects, where other tests fail to detect, the effects of brain lesions, drugs, radiation, and deprivation. Because of its demonstrated utility, the delayed-response problem, introduced by Walter S. Hunter one-half century ago, still is a favorite laboratory tool of comparative psychologists and neurophysiologists.

The interest of comparative psychologists is understandable. Hunter (1913) originally presented the problem as one involving a response to a discriminative stimulus not physically present at the time of the response and therefore requiring for its solution the capacity for symbolic, or representative, processes. Considered at that time to be the first real test of "higher-order capacities," the delayed-response problem uniquely appealed to comparative psychologists as a method for differentiating species along a phylogenetic continuum.

The pioneering effort of Jacobsen (1936) established the delayed-response problem as a behavioral test sensitive to neurophysiological insult. That he tentatively assigned to the frontal lobes "immediate memory" functions is irrelevant. His contribution gave impetus to the subsequent concerted and productive efforts to specify brain mechanisms subserving the higher-order processes.

It is only natural that 50 years of research should produce a prodigious amount of data and many analyses or interpretations of the true nature of the delayed-response problem. Most of the analyses, however, have been restricted to particular observations in quite specific situations. To my knowledge, there has not been any major attempt to summarize and analyze the accumulating diverse results of current behavioral research using the delayed-response problem. It therefore seems singularly appropriate to do so here.

A. Description (Direct Method)

Since its entry into the repertoire of laboratory tools, many variations of delayed-response testing have flowered but only one has flourished. This most common version is referred to as the *direct method* and will be described as it takes place in the ubiquitous Wisconsin General Test Apparatus (WGTA; see Chapter 1 by Meyer *et al.*). Figure 1 shows an adult rhesus macaque (*Macaca mulatta*) making a successful delayed response in a WGTA.

The delayed-response problem is presented in four phases: the baiting, covering, delay, and response phases. A trial begins with the *baiting phase,* which starts with the raising of the opaque screen between experimenter and subject. Beyond the reach of the animal, the typical test tray contains two empty foodwells behind which are located two identical objects. The experimenter in some manner attracts the attention of the subject, and in plain view and with obvious movements places the bait in one of the exposed foodwells. The *covering phase* follows immediately as the experimenter covers both foodwells simultaneously with the two identical objects. The *delay phase* consists quite simply of the experimenter withholding the tray beyond the animal's reach for a predetermined interval. At the end of this interval the *response phase* begins as the tray is pushed forward, allowing the subject to respond. A trial, of course, is terminated when the subject displaces one of the objects; this noncorrection procedure is now standard. The subject's performance is expressed in terms of the percentage of correct responses made over some specified number of trials.

These, then, are the essential operations. Ostensibly the delayed-response problem is a simple one, offering little difficulty to virtually any organism. Length of the delay interval is obviously an important variable, but, as we shall see, it is far from being the only variable that determines proficient performance of primates.

B. Major Research Efforts

1. PHYLOGENETIC DIFFERENTIATION

The first 20 or more years of research using the delayed-response problem were devoted almost exclusively to determining the maximum delay successfully tolerated by various species. The procedure consisted

FIG. 1. Rhesus monkey retrieving reward after displacing correct covering object in a typical Wisconsin General Test Apparatus. (Photograph by Fred Sponholz.)

essentially of running each animal under a particular delay to a criterion level of performance, repeating with a longer delay interval, and reporting the maximum delay duration at which criterion performance was attained. This immediate application of the delayed-response problem, i.e., phylogenetic differentiation, seemed reasonable in view of the early suggestion that the problem measured the so-called higher-order processes of memory or representative processes. An unfortunate aspect of much of this early research is the lack of attention given to the situational determinants of delayed-response performance. As a consequence of this failure to establish adequate control of the behavior in question, wide ranges of maximum delays were reported for the same species. We now know from later research that these apparent discrepancies are easily attributable to inadvertent procedural variations. Similarly, some procedures devised by recent experimenters have enabled nonprimates to perform well at delays once believed beyond their abilities (see Chapter 7 by Warren).

Despite disagreement about absolute maximum delays tolerated by a given species, the more extensive and well-controlled early comparative studies did concur in the general observations that primates higher on the phylogenetic ladder can tolerate longer delays under comparable conditions (Harlow & Bromer, 1939; Harlow *et al.*, 1932; Tinklepaugh, 1932). These studies represent but a fragment of the early comparative research on primates, and the reader seriously interested in the delayed-response problem is urged to read Tinklepaugh (1928) for a good historical introduction, Heron (1951) for nonprimate phylogenetic comparisons and early analyses of the problem, and finally Harlow (1951) for a survey of early primate literature.

Recent experimenters, inheriting the benefits of standardized methods, have verified the general correlation between phylogenetic status and performance on the delayed-response problem without primary regard to absolute limits of delay. The modern procedure is to select a number of delay intervals, randomly present these delays throughout a daily test session, test until some asymptotic level of performance is reached, and simply plot the percentage of correct responses made at each delay interval. Figure 2 is a composite graph of generalized performance functions of representative mammalian species as reported in a number of independent comparative studies. Because it is axiomatic that standardized procedures are not standard from laboratory to laboratory, the reader is urged not to interpret the performance of any species in terms of absolute values. The particular researches from which the graph was made only approximated comparable methods.

Even if it could be assumed that the data in Fig. 2 were obtained under identical conditions, the present emphasis would not be that performance is related to phylogenetic status, but rather that performance significantly deteriorates over so small a range of delay intervals for each species. These data argue against common sense, for if, indeed, the delayed response is primarily a simple test of recent memory, then a beast as bright as the rhesus monkey has obvious trouble remembering for as little as 20 seconds, and the cat has virtually forgotten during the same delay period. Intuitively, then, an animal's performance in the delayed-response problem must be influenced by variables other than the delay interval itself, and it is to the identification and elaboration of these determining variables that current research is directed. The advantage of such an approach is that the exact nature of the problem becomes exposed as we are able to specify the controlling variables. Moreover, only when we are able to control these determining variables will we be able to have a truly comparative research effort.

2. DETECTION OF BRAIN DAMAGE

The delayed-response problem is particularly sensitive to damage to the central nervous system, especially the frontal lobes. In his classic monograph, Jacobsen (1936) demonstrated the total loss of delayed-

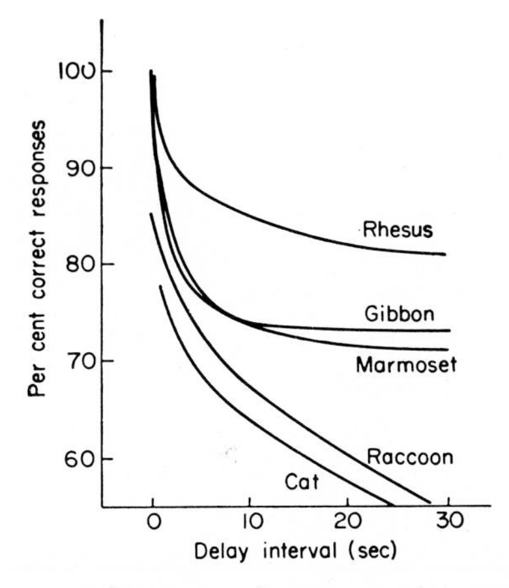

FIG. 2. Delayed-response performance of various species under similar test conditions. Data are redrawn from the following sources: gibbon from Berkson (1962); rhesus, marmoset, and cat from Meyers *et al.* (1962); raccoon from Michels and Brown (1959). Using a slightly different measure, the raccoons' performance was similar to the marmosets'.

response performance in several Old World monkeys with bilateral frontal-lobe damage, and he attributed the behavioral deficit to the loss of an "immediate memory" capacity. Subsequent neurophysiological investigations have made four consistent major attempts: (1) to refute Jacobsen's "immediate memory" hypothesis; (2) to determine the minimum effective lesion, hence delineate the essential neural mechanism that mediates this class of behavior; (3) to specify in greater detail the exact nature of the postoperative behavioral changes; and (4) to offer the most parsimonious explanation of the causes of performance decrement on delayed-response problems.

The first of these attempts is somewhat amusing because, when he chose the term "immediate memory," Jacobsen added the following footnote, which is as appropriate today as it was in 1936: "In using the term immediate memory to designate the defect that follows injury to the frontal areas we do so with little assurance that it is either sufficiently inclusive or descriptively adequate for the phenomenon in point. In some respects recognition memory and recall appear to be better suited. It is obvious that use of any of these terms adds little to our understanding of the essential physiological and psychological problems beyond a comfortable feeling of familiarity. For the present operational definition of the functions involved may be a more satisfactory procedure" (Jacobsen, 1936, p. 53). Despite this admonition the "immediate memory" hypothesis continues to be challenged.

However, there is no challenging the fact that frontal lesions produce delayed-response decrement. The search for the critical lesion, or the isolation of the essential neural mechanisms, appears to be fruitful. That a unilateral lesion is not sufficient to produce a response decrement was shown originally by Jacobsen and more recently by Meyer *et al.* (1951). Frontal and posterior mechanisms apparently mediate completely different behaviors. Harlow *et al.* (1952) demonstrated the typical decrement in delayed-response performance by rhesus monkeys with frontal lesions without a concomitant decrement in object-discrimination performance. Conversely, subjects with posterior lesions showed no loss of delayed-response capacity, but they performed poorly on object-discrimination problems. Lesions in the caudate nucleus also impair delayed-response performance, but less than do frontal lesions (Battig *et al.*, 1960; Dean & Davis, 1959). The most successful attempt to define the critical lesion was that of Mishkin (1957), who concluded that the midlateral frontal cortex contains mechanisms essential for delayed response in the monkey. Neurophysiological research pertaining to delayed-response problems, and neurophysiological speculations generated from such research, have been discussed in three excellent and current articles (Mishkin, 1957; Pribram, 1960; Rosvold, 1959).

Brain-damage research is obviously important to neurophysiology, but for the more behavioristically-inclined experimenter this research is important only if it helps explain performance of normal animals on the delayed-response problem. That it does will be emphasized in the following sections by showing essentially that the lesions themselves produce particular sets of behavior patterns that militate against solution of the problem. These behavior patterns, exaggerated in lesioned animals, may then be considered to reflect some of the variables that determine the responses of normal animals.

Next to be presented, therefore, is an analysis of the delayed-response problem followed by an examination, in terms of this analysis, of the major determining variables consistently demonstrated in research with both normal and brain-damaged primates.

II. AN ANALYSIS OF THE DELAYED-RESPONSE PROBLEM

As mentioned above, there have been many "explanations" of response deficits, hence many interpretations of the delayed-response problem. The more global interpretations were offered before the enumeration of the many variables that influence performance, especially performance of primates. The recent trend has been to offer more provincial interpretations attempting to explain response deficits in rather restricted experimental situations. While this strategy is perhaps prudent, the result has been a confusing proliferation of notions differing more in terminology than in content. The reason for this diversity of interpretation, especially in view of the growing ability to control performance, is not at all apparent; what is apparent is the diversity. A recent article, concerned with ontogenetic development of delayed-response capacity, offers a significant indication of current disagreement: "Psychologists have often been at a loss as to how to classify the delayed response. It has been variously considered as a learning problem, attention problem, memory problem, and thinking problem" (Harlow *et al.*, 1960, p. 120).

It is manifestly impossible, therefore, to discuss the many variables known to influence delayed-response performance in terms of the overly restrictive and diverse analyses typically offered with each experiment. Rather, for expository convenience and theoretical parsimony, and for any insights it may provide, I offer the following analysis in an attempt to assimilate existing analyses and to provide a unifying framework within which to interpret these determining variables.

There are, of course, boundary conditions within which, and only within which, any analysis is deemed to be appropriate. The conditions for this analysis are as follows. The discussion is restricted to an animal that is presumed to be completely "adapted" to the WGTA and associated testing procedures. Such an animal, if presented the test tray with exposed food in the foodwell, will look at, reach for, and consume the food immediately. The adapted animal also has all the behavioral mechanisms necessary for displacing an object that covers food. If the tray is moved within reach of the animal immediately after covering a single baited foodwell, there is no task for the adapted subject; it quickly and successfully displaces the covering object, and retrieves and consumes the food. This behavior of the adapted subject is, of course, learned during adaptation; but for an analysis of delayed response, it is superfluous to explain the strength of this behavior. The delayed-response problem, as a behavioral test, is used only after the above behavior exists.

Granted the above boundary conditions, the central concept in the analysis proposed here is that *the delayed-response problem is an intratrial performance task in which the correct instrumental response is the overt terminal response of an orienting-response chain initiated totally, accurately, and immediately by an observing response made at the beginning of each trial.* It is assumed that the exposed food elicits an observing response directed to the baited position and all cues associated with that position. The sampled cues may include the spatial stimuli unique to the particular side of the apparatus, distinctive object cues if any exist, and any proprioceptive cues occurring at the time of the observing response. It is assumed further that the observing response in turn initiates an orienting-response chain, directed to the baited position, the strength of which may be operationally defined by the overt terminal instrumental response made under a zero-second delay interval. But because the instrumental response is not allowed when a delay is introduced, it must be assumed that some covert component of this orienting response is elicited and that this covert response tendency will manifest itself in whatever overt orienting behavior is possible during the delay phase. This behavior may include bodily orientation such as positioning in front of the baited side, continued or repeated visual orientation, or complex responses that lead to the correct position.

The above statements are the essence of the position taken in this chapter. Certain corollaries are obviously needed to complete the position, but these are best introduced during the discussion of each phase of the delayed-response problem. What has been offered is sufficient to reveal the analytical problem. First, one must specify the stimulus conditions that influence the initiation and maintenance of the orienting-response chain. Second, one must specify the situational and organismic determinants of responses that compete with the orienting-response chain and ultimately reduce the probability of an instrumental response to the correct position. Though intratrial determinants of behavior will be emphasized, intertrial determinants also will be discussed.

It is, of course, quite easy to postulate behavioral tendencies so long as one is not constrained by empirical data. What is obviously necessary now is to consider the research on situational and organismic variables that are in fact related to performance, and to see to what extent this analysis helps make sense out of the various observations.

It must be noted first, however, that there is nothing essentially new in the above analysis. The interpretation of delayed response as a performance problem was suggested by Stanley and Jaynes (1949). Certainly much attention has been given to the role of "observing responses" recently, but the particular use of the term in this chapter was first suggested by Polidora in an analysis (Polidora & Fletcher, 1964) of the stimulus-response spatial-contiguity issue (see Chapter 1 by Meyer *et al.*). The fundamental notion of covert, implicit, or anticipatory responses comes directly from the early writings of Cowles (1940, 1941a, 1941b).

Finally, intertrial effects were stressed by Cowles (1941b) and by Harlow (1951).

III. DETERMINING VARIABLES

A. Baiting Phase

The initial and hence the critical events are the observing response elicited by the placement of the bait, and the orienting-response chain, overt or covert, initiated by the observing response. The following variables influence delayed-response performance, according to this analysis, because they affect these initial reactions.

1. Hyperactivity and Distractibility

A number of studies, in diverse research areas, consistently correlate one observation with decreased delayed-response performance: increased general activity and apparent distractibility. At the time of baiting, the tester can only try to get the animal to observe the baiting operation or the bait. While experience and skill certainly help in deciding when the animal has observed, any tester will admit that there are instances when this decision is difficult to make. In terms of the present analysis, it is assumed that hyperactivity and distractibility are response tendencies which interfere with the initial response of observing the bait and with the maintenance of the orienting-response chain that is elicited if the bait is observed (see Section III, C, 3).

Brain-lesion research offers the best single line of evidence supporting this assumption. Primates with frontal lesions, for example, are intensely hyperactive and distractible, and emit ritualistic, stereotyped response patterns. Hyperactivity *per se,* however, does not appear to be the critical factor, for animals with unilateral frontal lesions are similarly hyperactive without the concomitant performance decrement. The critical factor is apparently the related distractibility of the subjects, which makes them unable to observe or fixate the baiting of the foodwell. Meyer *et al.* (1951) reported that unilateral rhesus monkeys, though hyperactive, could be made to look at the bait, but bilateral animals did not so attend.

While much research indicates interference with visual orientation, there are no data to indicate, nor has anyone suggested, that the visual system itself produces deficient performance (see, for example, Glickstein *et al.,* 1963; Pribram, 1955).

Because certain cortical lesions make animals hyperactive and distractible, administration of drugs that affect activity might influence delayed-response performance of lesioned animals. This is easy to do and has been done repeatedly. Wade (1947), for example, trained rhesus monkeys to perform successfully at delays of up to 35 seconds and then removed the frontal lobes. After the operation, the monkeys were hyperactive and distractible and performed at chance level, even at delays of 5 seconds. But when they were given sedatives, they were less distractible and performed well at delays of up to 25 seconds. Only one subject retained delayed-response performance after sedation treatments were stopped.

A similar study by Dean and Davis (1959) indicates the startling ease with which delayed-response performance of monkeys may be manipulated. Rhesus monkeys, preoperatively trained to a high level of performance, were subjected to caudate lesions. Their postoperative performances on the same problems were typically poor, but phenidylacetate and reserpine restored preoperative performance levels. Cessation of drug treatments was followed by return by all animals to the poor postoperative level, indicating that the successful performance on delayed-response problems under the influence of drugs was not retained by the animals.

In view of these researches alone it is manifestly untenable to consider the delayed-response problem to be a test of retention capacity. An eminently more reasonable position is that certain drugs reduce hyperactivity and distractibility, thus increasing the probability of a sufficient observing response or of maintaining orientation during the delay (Section III, C, 3).

Additional support of a distractibility notion comes from radiation research (see Chapter 14 by Davis in Volume II). In an extensive research program on the harmful behavioral effects of radiation, Harlow and Moon (1956) compared irradiated and control rhesus monkeys on the delayed-response problem and found superior performance not in the control group but in irradiated animals. Conceiving of delayed-response performance in terms of "representative processes" leads to the conclusion that radiation improves the neural mechanisms serving such processes! The observable general effects of radiation suggest a more acceptable explanation. Irradiated animals are typically less active than nonirradiated animals. As hyperactivity is associated with performance deficits, hypoactivity might be associated with improved performance. Within a certain range, radiation effects are similar to those of mild sedatives (McDowell & Brown, 1958). Moreover, the results of a study designed specifically to investigate distractibility in irradiated rhesus monkeys led McDowell (1958) to conclude that radiation reduces distractibility. Similarly, in a recent survey of radiation research, Furchtgott (1963) attributes the general superiority of irradiated animals to decreased distractibility and narrowed attention. These conclusions are perfectly consonant with the observing-response viewpoint.

Finally, at least three recent investigations indicate that female rhesus monkeys perform better than males on the delayed-response problem. Blomquist (1960), using an indirect method (Section V, A), found this sex difference at delays of 6 and 12 seconds. McDowell *et al.* (1960), in an intensive study of 21 females and 20 males matched for age, found the same sex difference at a 10-second delay and in addition provided evidence for an explanation. They tested all animals for "distractibility" and concluded that the females were less distractible. McDowell *et al.* (1962) confirmed these findings in a study of previously-irradiated rhesus monkeys.

The above performances on delayed-response problems—loss of ability following certain cortical lesions, restoration of ability following administration of drugs, improved performance of irradiated monkeys, and the superiority of female monkeys—may readily be subsumed under, and attributed to, the performance factor of hyperactivity and distractibility. This performance factor presumably interferes with a sufficient observing response and ultimately reduces the probability of a correct instrumental response.

2. INCENTIVE VALUE AND DEPRIVATION

A sufficient response of observing the bait immediately initiates a chain of covert responses, which terminates in the instrumental response to the bait. These covert responses, like all responses, must occur with some degree of intensity, and it is assumed that the intensity of the covert response chain is a direct function of incentive value and the motivational state of the organism. Moreover, it is necessary to assume that greater intensity increases the subject's ability to maintain an extended covert response chain, hence a higher probability of a correct overt instrumental response after the delay interval. Thus, performance is a function of motivation, but the explanation differs considerably from the drive-reduction interpretation in that the effect of motivational variables on delayed-response performance is assumed to occur before, rather than after, the instrumental response occurs.

Motivational variables were explored in early research. Nissen and Elder (1935) showed that increasing the amount of banana increased the limit of delays for chimpanzees (*Pan*). Two years later, Cowles and Nissen (1937) cleverly isolated the effect of the incentive. Chimpanzees were shown either a large or small amount of food before the delay but actually received a constant small amount following the delay. While the amount of food obtained was therefore held constant for both incentive conditions, performance was better when more food was shown before the delay. Clearly the effect of amount of food took place before the instrumental response was made. In addition to quantitative effects, quality of food was shown to influence performance. Investigating various Old World primates, Maslow and Groshong (1934) found better performance associated with preferred foods.

Recent investigations have confirmed the effect of motivational variables under more standardized conditions. Pribram (1950) demonstrated that food deprivation improved the performance of lobotomized chacma baboons (*Papio comatus*), and he concluded that poor performance is probably attributable to variables operating at or near the presentation of food. Over a range of 2 to 50 hours of food deprivation, Gross (1963) found delayed-response performance of rhesus monkeys to be a function of deprivation, but no such effect was found in a delayed-alternation problem (see Chapter 5 by French). According to this analysis, one would not expect an effect on delayed-alternation performance because no food is shown, hence no orienting response is elicited, before the intertrial delay. Gross suggested that deprivation affects cue value, not reward value, in the delayed-response problem.

Finally, Meyer and Harlow (1952), using 1, 2, 3, or 4 pieces of food with normal rhesus monkeys, found that more food led to better performance, and Berkson (1962) demonstrated that gibbons (*Hylobates lar*) perform better for preferred foods.

In summary, motivation affects performance, according to this analysis, by varying the initial strength and maintenance of the covert response chain, thereby increasing the probability of the end result of that chain, the correct instrumental response.

The analysis thus far has emphasized the initial observation and a presumed covert response chain elicited by the bait and directed to the foodwell site. A recent study by French (1959) provides data which are most amenable to this analysis. French pretrained adult squirrel monkeys (*Saimiri sciureus*) on a standard delayed-response problem with a 10-second delay. A 5-second delay was then used in four test conditions. In condition A, baiting and covering phases were normal, but, during the delay phase, a barrier was placed over the negative foodwell, which thus was inaccessible during the response phase. Condition B was the same except that the positive foodwell was made inaccessible, and a correct response was defined as not responding to the negative object. In condition C, the negative foodwell was hidden during the baiting phase; both foodwells were accessible during the response phase. Similarly, in condition D, the positive foodwell was hidden during baiting (the monkey was shown the unbaited negative foodwell), and response was permitted to either site during the response phase. The ease with which the subjects solved conditions A and C supports the assumption that the response of observing the bait elicited a tendency to respond only to the baited foodwell. The unbaited foodwell site is presumably irrelevant in the delayed-response problem. (French also reported that during the baiting and delay phases under conditions A and C, the squirrel monkeys tended to orient toward the baited foodwell. This is considered to be the initial overt manifestation of the implicit response chain.) Conditions B and D were difficult. Under condition B the animals could not make the instrumental response to the baited foodwell. No response tendency toward the negative foodwell should have been established within the trial, but all prior training had involved reward of an instrumental response; so the competing response tendencies in this situation are obvious. In condition D, on the other hand, there was no response of observing the bait, hence no elicitation of an instrumental response chain. If observing an empty foodwell initiated an instrumental response chain of not responding to that foodwell, then this condition would have produced essentially the same results as condition C. That condition D was more difficult than condition C implies at least that a stronger covert response chain is elicited by food than by an empty foodwell. Thus French's data are explained easily by the present arguments.

B. Covering Phase

It seems plausible to assume that the covert response chain elicited by and directed to the baited foodwell remains unaffected by covering the foodwells. But since responses are made to stimuli, it is important to consider the change in the stimulus situation. The observing response elicited by the bait suddenly includes visual sampling of all discriminative cues provided by the covering object in addition to the discriminative cues provided by parts of the apparatus spatially contiguous to the baited foodwell. Typically the covering objects are as identical as possible, but it does not follow that there are no discriminable cues associated with each foodwell. It is assumed that the more discriminably different each foodwell site is from the other, the more stimulus-specific will be the elicited orienting-response chain, the greater will be its resistance to interference by general competing response tendencies, hence the greater will be the probability of a correct response. There are two obvious methods for producing greater discriminability of the foodwell sites: increased spatial separation and the use of discriminably different objects.

1. SPATIAL SEPARATION OF FOODWELLS

Before the WGTA became a standard laboratory instrument, Nissen and his associates explored the effect of spatial separation of food receptacles on delayed-response performance of chimpanzees. Carpenter and Nissen (1934) found that greater distances between boxes were associated with better performance and suggested that poor delayed-response performance was the result of interfering with the reinstatement of the visual field as given during the presentation. This effect was replicated 2 years later (Nissen *et al.*, 1936) and experimentally analyzed still later in a study designed to distinguish between the effects of spatial separation during the baiting phase and the response phase (Harrison & Nissen, 1941). Harrison and Nissen baited food containers under various spatial separations, but during the delay phase an opaque screen was lowered and the containers were shifted to different separations. Better performance was associated with greater spatial separation; moreover, the chimpanzees responded to both absolute and relative positions. But most important is the finding that spatial separation during the baiting phase was more effective than during the response phase.

The data from these experiments, and from the last in particular, support the conclusion that the orienting responses elicited before the delay interval are directed to a point in space and that increased spatial separation between alternative positions adds distinctiveness to those positions. It matters little whether one chooses to emphasize visual cues or proprioceptive cues. Certainly both exist.

2. Similarity of Objects

The obvious way to increase the distinctiveness of the foodwell sites is to cover them with dissimilar rather than identical stimuli. Because these added cues are incorporated into a more stimulus-bound orienting-response chain, the prediction must be that performance will be better. Research in the WGTA has demonstrated this to be the case for normal rhesus monkeys (Braun, 1952; Meyer & Harlow, 1952) and for raccoons (Michels & Brown, 1959). Use of distinctive cues on doors improves even the performance of rats on the delayed-response problem (Bliss, 1960).

That animals respond better to discriminably different baited positions should surprise no one and is predictable according to virtually any interpretation of the delayed-response problem. But what happens if the objects are moved during the delay phase? Does one predict a response to the covering stimulus or the baited position? Early researchers addressed themselves to this question and found that chimpanzees responded primarily on the basis of position (Nissen & Harrison, 1941; Nissen *et al.*, 1938; Yerkes & Yerkes, 1928). Consistent verification of this response tendency led Nissen *et al.* (1938) to conclude that the delayed-response problem requires an available symbolic mechanism which exists for spatial but not for nonspatial factors in the chimpanzee. This interpretation stemmed from an attempt to analyze delayed response as a "learning-retention" problem.

According to the present analysis, however, these results are predicted simply because the orienting-response chain is directed to the baited position and only incidentally to the stimulus object, which is visually sampled *after* the response chain is elicited. Of course if animals are given special training in which the object is moved but always covers food, then this problem approaches a two-trial discrimination problem (an implicitly-reinforced baiting-phase "trial" and a normally-reinforced response-phase "trial") and subjects could be expected to solve these two problems with equal facility. Presenting the above reasoning, Hayes and Thompson (1953) tested experimentally sophisticated chimpanzees on nonspatial delayed-response problems (10-, 30-, and 60-second delays) with unique objects on each trial and compared performance on this task to performance of the same animals on ordinary two-trial object-discrimination problems with 10-, 30-, and 60-second intertrial intervals. They concluded that nonspatial delayed-response performance was indeed comparable to two-trial discrimination performance so long as the subject attended to the baiting phase.

Recently Michels and Brown (1959) demonstrated the relative effects of added nonspatial cues on delayed-response performance of raccoons. After having trained their subjects extensively in the usual manner with delays up to 45 seconds, the experimenters arranged three "cue" conditions to be administered randomly with delays of 5, 15, 20, and 30 seconds, with the opaque screen down during the delay phase. In condition A (standard), two identical black cylinders covered the foodwells; in condition B, a black circle always covered the right foodwell and a white cross always covered the left; condition C was the same as B except that during the delay the stimulus objects were switched. Condition B produced performance which was better than that under A which was better than under C. These results are perfectly predictable from the proposed analysis.

The importance of the distinctiveness of all predelay cues is receiving increased attention and emphasis. For example, Campbell and Harlow (1945) attributed delayed-response deficit to inability to fixate and attend, failure to associate the position of the object with the implicit reinforcement, and inability to discriminate these cues from the rest of the situation. More recent research has indicated that the distinctiveness of cues is a critical variable in the delayed-response problem (Mishkin & Pribram, 1956; Pribram & Mishkin, 1956). After surveying this research, Pribram concluded that "frontal lesions interfere with whatever occurs at the time the delay task is set in the delayed reaction type of experiment, not with the process of recall (at the time response is allowed) per se" (Pribram, 1960, p. 20). The analysis presented in this paper continues and extends the current emphasis on predelay events by assuming that the task is completely set before the delay and that subsequent factors, for the most part, simply produce competing response tendencies which militate against successful completion of the task.

C. Delay Phase

According to this analysis, maintaining the orienting-response chain during the delay is crucial for the successful instrumental response at the end of the delay. The chain is maintained to the extent that the initial element (visual orientation) is maintained, or to the extent that some portion of the covert response chain results in some overt behavior directed to the correct position. Variables that operate during the delay phase to produce response tendencies incompatible with these orienting-response tendencies indirectly reduce the probability of a correct instrumental response occurring later during the response phase. We consider first the various orienting behaviors and then variables that presumably affect performance by disrupting the covert or overt orienting-response chain.

1. Orienting Behaviors

a. Bodily orientation. Early in the history of the delayed-response problem, psychologists considered possible "mnemonic devices" or ways in which animals could "bridge the delay interval." An obvious first answer is the use of gross bodily orientations. Enough research was devoted to this proposition to show that bodily orientation is not a *necessary* condition for solution of the delayed-response problem. Yet because these orienting behaviors do so often occur, they must be considered as at least a partially sufficient condition.

Primates exhibit a wide range of behaviors during the delay period, not the least of which are behaviors easily classified as gross bodily orientation. Quite often a monkey will simply position himself in front of the baited foodwell and remain in that position throughout the delay. Or an animal may begin an intense series of backward flips, but on the side of the cage corresponding to the baited foodwell. These orienting behaviors occur with such frequency that they cannot be disregarded in an analysis of the delayed-response problem. Indeed, if one accepts the assumption of an elicited orienting-response chain, then it seems entirely plausible to expect any covert response tendency to initiate an overt response directed to the baited foodwell even though the terminal instrumental response is not yet allowed. In fact, if the intensity of the orienting-response chain is a function of deprivation or incentive value, then the incidence, and perhaps form, of overt orienting behavior during the delay period should also be a function of these motivational variables. Unfortunately, there is no evidence to corroborate or refute this assertion. Nor is there likely to be such evidence in view of the current emphasis on collecting and reporting only quantitative data (number of correct responses) appropriate for elaborate statistical analyses.

Systematic overt orienting behaviors do not invariably occur, and the situational and organismic determinants of these behaviors remain elusive. There are suggestions, however, that New World monkeys may depend on these behaviors for solution of the problem more than do Old World monkeys. For example, in a single study Miles (1957) compared delayed-response performance of marmosets and rhesus macaques, and he suggested that body orientation (positioning in front of the baited foodwell) was necessary for marmosets but not for rhesus monkeys. For squirrel monkeys also, gross bodily orientation is associated with successful performance (French, 1959).

b. Visual Orientation. A frequently less obvious form of orienting behavior is visual orientation. Some animals will sit throughout the delay phase maintaining uninterrupted visual orientation to the baited position, and for such animals errors are rare. Other subjects may vigorously pace back and forth in the restraining cage while continuing their visual orientation. But many monkeys, regardless of their gross behavior during the delay period, glance almost furtively at the baited position. Whether continued or occasional, these visual orienting responses can obviously help to bridge the delay interval.

c. Covert Orientation. Overt orienting responses may not be necessary for solution of the delayed-response problem by some primates. Certainly at the human level one must consider covert orienting behavior in the form of verbal responses. Spiker (1956), for example, demonstrated better delayed-response performance by children who had learned distinctive names for stimuli than by children who had merely been given discrimination training on the same stimuli. Spiker suggested that having verbal names permits some form of representation of the absent stimulus during the delay. These results were later extended to mentally-retarded children by Barnett *et al.* (1959), who concurred with Spiker's interpretation. The analysis offered here would suggest that the verbal response is part of the covert response chain initiated by the observing response, and that maintaining this part of the chain by repeating it covertly during the delay serves to protect the integrity of the entire response sequence ending in the correct instrumental response.

2. VISIBILITY OF THE TEST TRAY

To increase the difficulty of the delayed-response problem, simply lower the opaque screen for the entire duration of the delay. This change in the stimulus situation increases errors by 50% in normal rhesus monkeys (Meyer & Harlow, 1952). The effect is not limited to normal animals. Rhesus monkeys with frontal lesions failed completely with the screen down for the *minimum possible* duration (Battig *et al.*, 1960).

Not only does performance drop drastically at short delay intervals with the screen down, but rhesus monkeys may remain at low performance levels for long periods even though they are capable of successful response after the same delays with the screen up (Cadell, 1963).

The obvious effect of lowering the screen is to completely disrupt visual orientation. The poor performance always found under this condition suggests that the entire orienting-response chain is weakened drastically as a result of disrupting the visual orienting component.

Severely disrupting, reliably demonstrable, and easily controllable, the operation of lowering the opaque screen has nevertheless been strangely devoid of the usual parametric investigations, and at least one interesting possibility exists. The standard procedure is to bait the foodwells, cover them with the objects, and lower the screen immediately. A skilled tester performs these operations quickly, and the effect may be that the orienting-response chain is not sufficiently initiated by so short an observing response. A prediction consistent with the present analysis is that, for a given delay period (e.g., 20 seconds), performance will be positively related to length of time allowed before the opaque screen is lowered. In fact, there may well be some critical interval beyond which the lowering of the screen has no effect on performance. Whatever the results, research along these lines will serve to differentiate further the relative influence of a maintained visual orienting response and any other overt or covert orienting responses made within the delay phase.

3. Hyperactivity and Distractibility

The discussion of activity and distractibility during the baiting phase (Section III, A, 1) is equally appropriate here. The integrity of the orienting-response chain will be inversely related to competing response tendencies, i.e., tendencies that force the subject to look away from the foodwell area or that interfere with the initiation or maintenance of some particular overt orienting-response chain leading to a correct response. Hyperactivity and distractibility can, of course, affect either or both elements in the chain.

Despite the emphasis placed on hyperactivity and distractibility of frontal primates since Jacobsen's original observations, few writers have made the distinction between, or have investigated, activity during the baiting phase as opposed to activity during the delay phase. Both must be important. Malmo (1942), testing a rhesus macaque and a sooty mangabey (*Cercocebus torquatus*), was careful to observe the activity during the delay interval. He reported that higher activity was associated with lower performance. Moreover, when darkness was maintained during the delay period, performance was better. This observation led Malmo to propose his retroactive-inhibition interpretation, which in essence states that frontal animals are more susceptible to interfering effects of extraneous stimuli.

There is no reason why increased activity or distractibility, exaggerated in the frontal subject, should not also determine to some extent the performance of normal monkeys. Indeed, my own observations in delayed-response testing convinced me that activity during the delay period was critical even in normal monkeys. This hypothesis was tested at the University of Wisconsin Primate Laboratory by arranging a photocell assembly across the rear section of the restraining cage and defining activity by the number of photobeam interruptions *during the actual delay intervals* only (Fletcher, 1964). Five adult rhesus macaques that had been trained with random presentations of 0-, 5-, and 10-second delays were tested for six sessions with randomized 0-, 5-, and 20-second delays. During these sessions, no errors were made under the 0-second-delay control condition, indicating that the baiting operation did in fact elicit the appropriate orienting response. However, responses under the delay conditions of 5 and 20 seconds were 91% and 84% correct. The rank-order correlation of the total number of errors and the total activity score made by each animal during all delay periods was almost perfect (rho = 0.975). These data, then, strongly support the conclusion that generalized activity, which is random with respect to the baited position, interferes with the maintenance of the elicited orienting chain directed to that position.

The preceding statement does not imply that generalized hyperactivity will invariably produce behavior incompatible with correct delayed-response performance. In any situation involving two competing response tendencies, there are three possible results: the first may inhibit the second; the reverse may occur; or both may be incorporated into a behavior pattern compatible with each tendency. Once again primates with frontal lesions, because of their exaggerated behavior patterns, provide excellent illustrations.

First, in view of the general observation that the hyperactive frontal animal is a notoriously poor performer on the delayed-response problem, one may assume that the resultant response tendency for generalized activity (or the concomitant distractibility) is preemptive and disrupts any directive behavior patterns that might have been initiated. This seems to be true of squirrel monkeys, which normally rely heavily upon orienting behavior. Squirrel monkeys with frontal lesions were unable to maintain orienting responses and reportedly behaved as if the baiting event "did not instigate any continuing input" (Miles & Blomquist, 1960, p. 481). While in essential agreement, the present analysis suggests that an orienting-response chain may have been instigated, but hyperactivity prevented the manifestation of overt orienting behavior. In accord with this notion, frontal rhesus monkeys tested by French and Harlow (1962, Exp. 3) performed well at 0-second delays but not at 5- to 20-second delays.

Some frontal rhesus monkeys do perform successfully on delayed-response problems, and their success is often correlated with the maintenance of orienting behavior. For example, Orbach (1962) reported adequate performance by multilesioned young rhesus monkeys, each incidence of which was accompanied by positioning in front of the correct foodwell. Glickstein *et al.* (1963) reported that one lesioned rhesus monkey picked at the floor near the baited foodwell and frequently glanced at it during the delay. When this stereotyped behavior did not occur, performance fell to chance level. Another subject intermittently grasped the cage bar nearest the baited foodwell throughout the delay. "Successful performance was contingent on maintaining this directive behavior" (Glickstein *et al.*, 1963, p. 14). These data suggest that, under some (as yet unspecified) stimulus conditions, the hyperactivity of frontal rhesus monkeys may be inhibited by the overt orienting-response sequence directed to the baited foodwell.

The final possibility remains that the random hyperactivity pattern of the frontal primate may be so modulated by the orienting-response chain that the intense, gross-bodily-movement pattern itself becomes a directive behavior pattern. Pribram (1950) demonstrated that successful performance of lobotomized chacma baboons was correlated with motor patterns (tapping, pacing, etc.) that tended to force an approach to the correct position first. The degree of complexity of the directive response sequences possible was amply demonstrated in a study by Orbach and Fischer (1959). These investigators observed the usual intense circling behavior following bilateral frontal lesions in rhesus monkeys, but they were able to detect *clockwise* circling when the right foodwell was baited, *counterclockwise* circling when the left foodwell was baited.

It must be emphasized that the typically-reported description of the frontal primate is that of generalized, random hyperactivity and distractibility. The research reported here, however, attests to our severe inability to specify the situational or organismic determinants of these behavior patterns. Once again, our knowledge concerning these determinants will increase only when we have detailed descriptions of behavior occurring in the delayed-response problem, descriptions which are so often neglected in psychological journals.

4. Sequence of Presentation of Delay Intervals

Whether by design or chance, the evolution of delayed-response testing has resulted in the "standard" procedure of randomly presenting all of the delay intervals under investigation within a given session. Rapidly-accumulating evidence suggests that this random presentation of the different delays, rather than the delay duration itself, may be a variable that lowers performance.

Testing normal and frontal rhesus monkeys for discrimination learning under intertrial intervals of 10, 30, 60, and 120 seconds, Riopelle and Churukian (1958) found that even the frontal subjects performed successfully following the long intervals. So a frontal monkey is able to "retain" a discriminative response for at least 2 minutes, and therefore the delay *per se* cannot be the critical variable in the delayed-response problem. This research supports an earlier finding of Finan (1942), who showed that if frontal sooty mangabeys are allowed to make a predelay response, then they can perform successfully at delay intervals up to 30 seconds. In a completely different context, Mishkin and Weiskrantz (1958) investigated the effect of delaying reward in a visual discrimination task performed by rhesus monkeys with frontal lesions. An 8-second delay between the response and reward produced poor performance in the frontal animals. However, if the delay were gradually introduced, there was no such impairment; therefore the intratrial delay was not a sufficient cause of poor performance. On the basis of these data, Mishkin and Weiskrantz compared the delay in a discrimination problem to the delay in the delayed-response problem. They concluded that the position of the delay, whether between the stimulus event and the response (delayed-response problem) or between the response and the reward (discrimination problem), is irrelevant, and nondelay variables should be emphasized.

The above research clearly argues against a time factor *per se*, and the last study suggests that whatever behaviors are necessary to bridge a delay may be established by training with progressively-increasing delay intervals. Current research indicates that this is the case in the delayed-response problem. Battig *et al.* (1960) found that gradually increasing the delay interval resulted in good performance of frontal rhesus monkeys at least up to a 5-second delay. By increasing the delay interval in 1-second steps, Glickstein *et al.* (1963) demonstrated rapid attainment of 90% correct response at a 15-second delay interval by rhesus monkeys. A most convincing demonstration of the effect of random presentation of delay intervals was reported by Riopelle (1959), who found that random presentation of delays of 15, 30, 45, 60, 75, and 90 seconds, while resulting in performance reliably above chance (about 60% for a five-choice situation), lowered performance at the 15-second delay about 20% below the level obtained previously at a 22-second interval (Section V, A). Riopelle concluded that "inclusion of many long-delay trials introduces factors, e.g., frustration and anger, which affect performance on the short delays" (Riopelle, 1959, p. 752). Gleitman *et al.* (1963) also reported more frustration when long delays precede short delays in training. They reported that with repeated long delays a rhesus monkey "loses interest," "stops paying attention," or "extinguishes an observing response."

Recent research, then, clearly indicates that rhesus monkeys, normal and brain-damaged, may be trained to perform relatively well on delayed-response problems by gradually increasing the delay interval. Random presentation of short and long delays, on the other hand, results in relatively poor performance. These empirical results also seem to be amenable to analysis within the present framework. Assume a random presentation schedule with a short delay. Any correct orienting-response chain will be reinforced at the end of the short delay. Assume that the next trial has a longer delay phase. The same response chain of a given length or duration is now not reinforced. Other response tendencies may then disrupt the chain, extinguishing the entire sequence, so performance at all delays deteriorates. On the other hand, an operant chain may easily be extended so long as the required increments are kept small. Thus, by gradually increasing the length of the delay phase, the mediating response chain is gradually extended in increments small enough that incompatible responses do not occur.

Whatever the ultimate explanation, the practical significance is clear. Should the experimenter be interested in maximum performance at a given delay, he should train his subject at a minimum delay and then gradually increase the delay duration. Should he be more concerned with relative effects and actually want to make the problem more difficult, then he should employ the more convenient and standard procedure of randomly presenting the various delay intervals.

D. Response Phase

Beginning with the assertion that a delayed-response trial is simply a performance trial, the analysis has thus far considered *intratrial* environmental events that presumably produce response tendencies which interfere with the appropriate response sequence elicited at the start of a particular trial. But behavior is always a function of prior experience; so we now must consider *intertrial* environmental events as they affect performance on the delayed-response problem.

1. Outcomes of Prior Responses

The outcomes of prior responses presumably could affect performance on a given trial. Yerkes and Yerkes (1928) reported that chimpanzees tended to respond to the box that contained food on the previous trial rather than to the box baited at the start of the particular trial. Consistent with the *Zeitgeist,* Yerkes and Yerkes interpreted these observations in terms of dominance of primary reinforcement over secondary reinforcement. A more recent and much more dramatic observation was reported by Wilson *et al.* (1963). Four rhesus monkeys, highly trained on the delayed-response problem, were subjected to bilateral frontal lesions, and retested postoperatively. On the first postoperative 5-second delay trial, 20 days after the last preoperative trial, each frontal animal responded to the foodwell to which the last rewarded response had been made! It must be noted, however, that for two animals the first-trial baiting occurred at the same foodwell last responded to, and so the responses of these animals were correct. Only two animals, then, responded in opposition to the baiting cue on the first trial. But regardless of the emphasis given to this particular observation, there is some evidence, and intuitive appeal, for the notion that intertrial effects may constitute a major determinant of response strength.

Though many writers have considered intertrial effects, the most emphatic expositions are given by Cowles (1941b) and Harlow (1951), both of whom interpreted the delayed-response problem in terms of a random series of discrimination reversals, a task difficult for any animal. Because there are only two positions to which the animal responds, and because the baited position is determined by a random sequence, the position-reversal element is obvious. On a given trial, a response to the left foodwell may be reinforced. On the next trial, if one is willing to assume that baiting constitutes some form of implicit or secondary reinforcement, the animal may be successful only so long as the secondarily-reinforced response tendency inhibits the primarily-reinforced response tendency established on the immediately preceding trial.

A theoretical position enjoying increasing currency is that the peculiar deficit of primates with frontal-lobe damage is an inability to inhibit response tendencies. In the most complete and detailed paper relevant to this viewpoint, Stanley and Jaynes (1949) posited their "cortical act-inhibition hypothesis," which essentially attributes to the frontal cortex the primary role of suppression or inhibition of entire response sequences or acts. Using delayed-response performance as an example, Stanley and Jaynes argued that getting food in one position will tend to make the animal respond to that position on the next trial. "But the normal animal in learning the problem must over-ride these effects of primary reinforcement by inhibition of what was learned on the previous trial. In frontal monkeys, who are relatively incapable of act-inhibition, this over-riding of the effects of primary reinforcement cannot occur and rigid learning of a position habit ensues on the basis of partial reinforcement" (Stanley & Jaynes, 1949, p. 29).

One of the most recent major advocates of inhibition theory is Mishkin, who has extended his inhibitory concepts to the delayed-response problem on the basis of an impressive line of research in discrimination learning. His views appear to agree completely with those of Stanley and Jaynes. Mishkin assumes that the delayed-response test elicits strong tendencies to respond to a particular spatial location. Frontal primates are presumably unable to suppress or inhibit this tendency. Therefore the position habits (repeated responses to a particular location) so often observed in frontal animals "are not simply the by-products of their inability to learn this test, but are the direct source of their failure" (Mishkin *et al.*, 1962, p. 181). Recent evidence (Battig *et al.*, 1962) has been interpreted as supporting inhibition theory and this may well represent a major trend in analysis of the delayed-response problem.

2. Distribution of Trials

The strength of intertrial effects must themselves be under the control of some variables. One such variable which has been investigated is that of frequency of trials within a session—the old question of massed versus distributed practice. Spaet and Harlow (1943, p. 432) briefly mentioned the performance of a yellow baboon (*Papio cynocephalus*) that failed with delays longer than 5 seconds when given 25 trials per day but attained 90% correct in a series of 25 delayed responses of 15 and 30 seconds when given only 1 trial per day. This superiority of distributed practice was recently confirmed by Gleitman *et al.* (1963), who tested rhesus monkeys under a massed condition of 20 trials per day and a distributed condition of 2 trials per day for 10 days.

The data on massed versus distributed practice have apparently supported inhibition theory. Assuming that food-reinforcement effects dissipate with time, inhibition theorists reason that the longer the interval between responses, the less the effects of previous primary reinforcements, hence the better should be the performance of frontal monkeys because "the situation does not demand so strong an inhibition of the previously reinforced response" (Stanley & Jaynes, 1949, p. 29). However, the same assumption immediately creates a logical difficulty for an inhibition theory which stresses the transfer effects of reinforced responses. For instance, one way to increase the time between a reinforced response and the occurrence of the subsequent response (hence dissipation of the effect) is to increase the delay interval while maintaining a constant intertrial interval. In such a case, inhibition theory, as currently stated (or interpreted by me), must predict that frontal animals will perform better following longer delays than following shorter delays! This prediction is absurd in view of the typical monotonic decline in performance as a function of delay interval for both normal and frontal primates.

A solution to this apparent dilemma is available and merely requires the specification of precisely when the competition of response tendencies (or failure of inhibition) is assumed to occur. The above prediction holds only when the theory assumes that response tendencies compete during the response phase. On the other hand, if inhibition theory assumes that response tendencies compete *before* the delay phase, then both delay duration and interresponse time (because they follow the competing events) are irrelevant, and only the intertrial interval is critical. This seems to be the most reasonable hypothesis in view of the data on massed versus distributed practice. But still lacking is a complete explication of the effects of intertrial interval, delay duration, and perhaps interresponse time. Clearly, the success of inhibition theory as a major explanatory position will depend on the results of parametric research devoted to this explication.

However, additional arguments may be advanced against inhibition theory as it applies to the delayed-response problem. First, the empirical evidence is not consistent. In alternation training, "frontal operatees do not appear to perseverate more than normals when they are rewarded" (Wilson, 1962, p. 703). Moreover, in delayed-response training, not all frontal animals develop the strong position habits which they are supposedly incapable of inhibiting. The conditions under which position habits occur are simply not yet known. Second, these position habits may actually be examples of the previously-discussed rigid, stereotyped behavior patterns, which tend to result in responses to one position rather than the other. While they must appear on a data sheet as a position habit, it seems a bit artificial to interpret these response patterns as being conditioned to a particular foodwell on the basis of an intermittent reinforcement schedule. Third, inhibition theorists have yet to incorporate into their system all of the variables demonstrated to affect performance. For example, how does lowering the screen increase a failure to inhibit an incorrect positional response tendency? Or in what way does the addition of trial-unique dissimilar stimuli improve the inhibition of an incorrect positional response tendency? And increased incentive value should, it would seem, facilitate acquisition of a position habit and make its inhibition more difficult, thus producing worse rather than better performance. Finally, it appears somewhat unwarranted to consider delayed-response failure in terms of failure to inhibit an intertrial incorrect response tendency before one can demonstrate that a correct response tendency has in fact been established within a trial. It is entirely possible, as proposed by the present analysis, that (a) the correct response tendency may not be elicited under certain conditions, or (b) once having been initiated, the correct tendency may be nullified by competing response tendencies resulting from certain intratrial events. Should either of these two intratrial results occur, the next-most-probable response tendency should be one established on the prior trial. We need to explain not why an animal fails to inhibit an incorrect response tendency, but rather why it fails to establish or maintain the correct response tendency. These may well be complementary questions.

IV. DELAYED RESPONSE AND DISCRIMINATION LEARNING CONTRASTED

The assertion that the delayed response is a *performance* problem is not meant to imply, nor should the reader infer, that no learning is involved. Because performance improves with practice, obviously some form of learning takes place. But I feel obliged to make explicit now the distinction between the learning involved in the delayed-response problem and the learning involved in a simultaneous discrimination problem.

Consider first an object-discrimination problem. On the first trial of a given problem there is no "correct response tendency," and the response is determined by position or object preference. What is important is the event following this and subsequent responses, for throughout a particular problem a response to only one specific, discriminably different, object will be consistently followed by reward. Whether one subscribes to an excitatory, inhibitory, or duoprocess theory, it is classically assumed that the events following a response somehow increase the capacity of a particular stimulus object to elicit the "correct" response on subsequent trials. Moreover, the "discriminative stimulus" (defined expressly as that specific cue which is consistently correlated with reward) is responded to, or is present at the time of the response. Once the discriminative stimulus reliably elicits the correct response, then that discrimination problem may be considered the limiting case of a delayed-response problem with zero-second delay but with the discriminative stimulus present at the time of response. The analytical problem, therefore, is one of specifying the conditions under which the arbitrarily-designated discriminative stimulus, on repeated presentations, acquires the capacity to elicit a correct response, an *intertrial* phenomenon.

A similar analysis of the delayed-response problem is not so easily managed. For example, one may interpret the delayed-response problem within a learning context by assuming that the discriminative stimulus to which the animal responds is the particular foodwell site, and all its contiguous stimuli, *provided that it was preceded by baiting*. However, the only *external* cues to which the animal responds are those spatially contiguous to the foodwell site, and these cues are not consistently related to reward from trial to trial; hence, according to the strict definition presented above, these cues cannot be termed discriminative stimuli. Therefore, some mnemonic representation of the preceding baiting phase must provide the distinctive cue, so that the discriminative stimulus to which the subject responds is a complex consisting of the spatial stimuli distinguished by the mnemonic representation. With practice the animal must learn to respond consistently to this relational cue.

While the above interpretation is a perfectly reasonable one, it clearly places the delayed-response problem back into the category of an "immediate memory" problem. This chapter, of course, has taken the entirely different, nonlearning approach as follows.

A delayed response involves no "discrimination learning" beyond the discrimination of bait from no bait. Rather, on trial 1 and every subsequent trial *independently*, the correct response chain is elicited completely and reliably by the baiting operation. The observing response, completed during the baiting phase, automatically establishes the stimulus (foodwell site and contiguous stimuli) to which the animal will respond, but it is completely artificial to refer to this stimulus as an ordinary "discriminative stimulus." That all necessary response tendencies are in fact established immediately is demonstrated by the typically errorless performance under a zero-second delay condition. Therefore, the analytical problem, according to this interpretation, is one of specifying the conditions under which a response, once initiated, is maintained within a given trial, an *intratrial* phenomenon.

Other writers, stressing the intratrial significance, have interpreted the delayed response as a learning-retention problem or one involving a predelay implicit reinforcement and a postdelay response. This writer obviously agrees with the essence of these interpretations but feels that the use of terms such as predelay "learning" or "implicit reinforcement" are unnecessary terms which simply state that the correct response is elicited at the beginning of the trial. In the delayed-response problem, therefore, the necessary discrimination is made and the response is initiated at the beginning of each trial, and the only learning that occurs with practice is the fixation of an adequate orienting-response chain which will terminate in a correct response.

V. VARIATIONS OF THE DELAYED-RESPONSE PROBLEM

A. Indirect Method

Actually introduced by Hunter (1913) before the direct method is a variant of the delayed-response problem referred to as the *indirect method.* Rather than starting a trial with food, this method uses a previously-established discriminative stimulus to indicate the rewarded position. For example, after training an animal to respond to an illuminated panel, the experimenter then tests his subject by illuminating one panel for some specified duration, terminating the illumination, enforcing a delay phase, and then permitting the subject to respond to the unlighted panels. An obvious disadvantage to this method is that it requires preliminary discrimination training to establish the capacity of the discriminative stimulus (lighted panel) to reliably elicit the correct response. With adequate pretraining, however, the indirect method and the direct method produce approximately equivalent results. There are certain advantages to the indirect method. The experimenter clearly has better control over the presentation of the discriminative stimulus. Duration, intensity, and quality, for example, may be manipulated with precision. The indirect method also reduces the uniqueness of the tester's manner of attracting the subject's attention during the baiting phase, but the assumption that an observing response has occurred must also be made with this method. Finally, the baiting phase and covering phase are combined into a "stimulus-presentation phase." The indirect method has not been so popular as the direct method, but the two recent studies which will be discussed may revive interest in the indirect method.

In an interesting series of experiments, Riopelle (1959) tested irradiated and normal rhesus monkeys under many experimental conditions. His modified WGTA included an opaque screen and a transparent screen, a tray with five boxes whose translucent front panels could be illuminated, and foodwells located behind the top edges of the panels well out of view of the subject. Each trial began with raising the opaque, but not the transparent, screen during the entire stimulus-presentation and delay phases. The first experiment examined the effects of stimulus duration (1, 6, or 11 seconds) and length of delay between stimulus termination and response (2, 7, or 12 seconds). It was anticipated that the longer the stimulus duration, the better the observation; hence the more probable would be a correct response. The opposite results were obtained, however, and Riopelle's interpretation is in accord with one suggested by the present analysis. The onset of the discriminative stimulus was sufficient to elicit the observing response and initiate the orienting-response chain directed to the illuminated panel. The remainder of the stimulus duration added only to the delay phase.

Later, with the same stimulus-presentation and delay durations, Riopelle compared the effect of illuminating all panels during the delay and response phases with the standard condition of all panels dark during these phases. This is equivalent to requiring a response with the discriminative stimulus *appearing* at all five positions versus requiring the usual response to an *absent* discriminative stimulus. Performance was, of course, worse when all panels were illuminated because, as Riopelle argued, the onset of illumination elicited observing responses which tended to initiate competing responses.

In a subsequent experiment, Riopelle verified the deleterious effect on performance of lowering the opaque screen during the delay (Section III, C, 2). He then increased delay intervals to 2, 12, and 22 seconds without lowering the opaque screen and obtained over 90% correct response for all intervals. In a final experiment, described in Section III, C, 4, six delay intervals were presented in random order.

In its typical application, the indirect method of delayed-response testing follows a pretraining phase during which the discriminative stimulus is learned. An ingenious procedural variation allowed Blomquist (1960) to assess simultaneously the acquisition of discriminative and delayed-response performances. His data argue unequivocally for a distinction between delayed-response performance and discrimination learning.

Blomquist used a WGTA with the usual opaque screen and a movable test tray with two foodwells. Two 4.5-inch-square boxes, with translucent white plastic front panels capable of being illuminated from within, could slide over the foodwells. The subject was required to push back the illuminated box. Blomquist investigated the effects of stimulus intensity (1 or 9 lamps), stimulus duration (1 or 3 seconds), intertrial interval (15 or 30 seconds), delay duration (0, 6, and 12 seconds)and sex (four female and four male 2-year-old rhesus monkeys were tested). His procedure was to push the tray partially forward, illuminate one box with a particular intensity and for a given duration, maintain the position of the tray for the delay duration, and then push the tray forward for the response phase. Under the 0-second delay condition the tray was pushed completely forward immediately, so this condition approximates an ordinary discrimination-learning situation with the possible exception that the stimulus duration of 1 or 3 seconds may have expired occasionally before a response was actually made. Ignoring this small discrepancy, one may consider the 0-second delay condition to be assessing the acquisition of a discriminative response and the 6- and 12-second delay conditions as measuring performance of the discriminative response in the absence of the discriminative stimulus.

Discrimination performance (0-second delay) reached and maintained a stable level of 97% correct responses from trials 504 to 720. During the same period, 84% and 72% correct responses at 6- and 12-second delays indicate the performance of delayed responses to the well-established, but absent, discriminative stimulus. Thus, these performance levels represent the terminal strength of the orienting-response chain elicited completely 6 or 12 seconds earlier by the discriminative stimulus. Clearly, discrimination and delayed-response performance are not equivalent.

Blomquist reports still more evidence which further differentiates discrimination learning from delayed response. He predictably found a statistically significant superiority of his female monkeys, but only under the delay conditions, not under the discrimination-learning condition.

Clearly, then, whatever is producing the sex difference, it does not affect the *learning* of a discriminative response, but it does affect the *performance* of the same discriminative response as measured in the delayed-response problem.

These two studies alone sufficiently indicate the flexibility and control inherent in the indirect method. The variable uniquely amenable to experimental manipulation is the discriminative stimulus and all its attributes. By controlling the level of prior discrimination learning, one should be able to control the level of delayed-response performance, if the present analysis is correct in its assumption that the response is completely elicited by the discriminative stimulus at the beginning of each trial. Because the indirect method is necessarily confounded with prior or concurrent discrimination learning and consequently must be sensitive to the same variables that control the acquisition and maintenance of a discriminative response, future research may well detect subtle differences between the indirect and direct method of testing the delayed response.

B. Rotating Test Tray

Behavior during the critical delay phase takes many forms, two of which are visual orientation to the baited position and positioning in front of the appropriate foodwell (Section III, B, 1). Either of these two behaviors can bridge the delay successfully and result in a correct response. But the typically compulsive experimentalist may be somewhat reluctant to offer to his subject the option of responding in either or both ways. A respectable degree of control over the subject's behavior is assured by making successful performance contingent upon one, but not the other, response. This may be achieved by moving the food containers or the entire test tray during the delay phase, and consequently making a correct response dependent upon continued visual orientation. This stratagem is, as usual, not new. Harlow (1932) shifted identical containers during a 10-second interval and demonstrated that some Old World monkeys could solve the problem, thus showing that bodily orientation was not a necessary condition for solution. Nissen *et al.* (1938) only briefly mentioned an interesting application of this principle. They reported using a small turntable on which either a black or a white food container was baited during rotation. Two chimpanzees were allowed to respond when the turntable stopped. When rotation was slow, performance was almost perfect, but when rotation was rapid enough to make visual following impossible, performance fell to chance. This unfortunately brief description certainly suggests the complete dependence on continued observing, or visual following, for the solution of a delayed-response problem involving two clearly distinctive food containers.

Recently Cadell (1963) made use of this technique. Investigating the effects of fornix damage in male rhesus monkeys, Cadell tested four control and eight experimental animals for 600 trials at 0- and 5-second delays with the opaque screen up, then 300 trials at the same intervals with the screen down, 300 additional retraining trials with the screen up, and finally 600 trials with the rotating test tray shown in Fig. 3. Standard

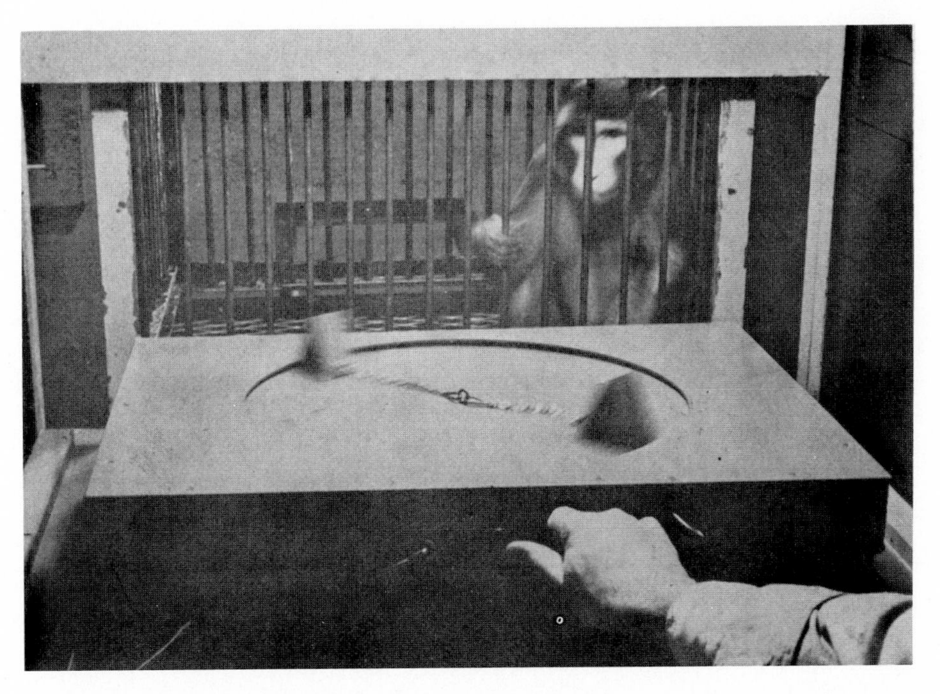

FIG. 3. A rotating test tray used in the delayed-response problem. Note the maintenance of visual orientation during the delay interval. (From Cadell, 1963; photograph by Fred Sponholz.)

baiting procedures were followed by covering both foodwells with the two identical red isosceles triangles used in previous testing. At the end of the covering phase the tray began either 1 or 1.5 rotations, stopping automatically in the same or the reversed position. These rotations took either 3.8 or 5.0 seconds and constituted the delay phase, after which the tray was pushed forward immediately. During this delay phase, while the tray was rotating, Cadell wisely recorded whether or not the monkey observed the stimuli.

Performance of both the experimental and control groups dropped reliably when the rotating tray was introduced. The critical analyses, however, required the classification of animals as "observers" and "nonobservers," since solution of the problem demanded continual visual orientation. Although this dichotomous classification did not differentiate the experimental and control groups, it did differentiate all animals with respect to performance. Performance of nonobservers fell completely to chance level and did not improve. Performance of observers dropped reliably from 85% to 68% correct responses, but improved back to 85% correct over the 600 trials. Although the classification criterion was obtained for each animal under the rotating-tray condition, Cadell reanalyzed the data from the standard condition and found the same statistically significant superiority of the observer animals. This is not at all surprising, because if visual orientation was elicited with different strengths among animals under the rotating-tray condition, then this same response was probably elicited similarly in the same animals under the standard condition. While not necessary for solution in that situation, visual orientation obviously can help bridge the delay, hence the better performance of the observer animals under the standard condition. Cadell's results demonstrate once more the desirability of recording and reporting as much of the animals' behavior as possible.

The unique advantage of the rotating tray is that it demands a specific form of orienting behavior, continual visual orientation. This test apparatus, then, should be most sensitive to experimental manipulations that produce response tendencies inimical to maintaining this particular orienting response. Moreover, this apparatus is easily converted to the indirect method (by using illuminated, flashing, or colored sliding or hinged lids) and may therefore provide the experimenter with the additional advantages of flexibility and control over stimulus presentation. Used with either method, the rotating tray may become a popular and powerful addition to the modern battery of research tools.

VI. CONCLUDING REMARKS

Hopefully, this examination of the delayed-response problem will stimulate research. Assuming that it will, I cannot end the discussion without imploring the experimenter to consider the following two essentials, the unfortunate omission of which has vitiated much of the existing research.

First, *it is mandatory to include a 0-second delay in all phases of the experiment.* Consider, for example, an experiment involving a brain lesion performed on a randomly-selected half of all animals trained to a criterion level at a 20-second delay interval. Assume that postoperative testing revealed reliably worse performance of the lesioned animals at the same delay. With all appropriate scientific caution the experimenter could infer that the lesion interfered with the ability to negotiate the same delay interval previously responded to successfully. Further theoretical speculations are irrelevant because the delay period has already been designated as the critical factor. Assume, on the other hand, that a 0-second delay was also included and an identical postoperative decrement appeared under this condition. Clearly some other inference must be entertained. In the extreme case, the lesion may have impaired the visual system to such an extent that the animals had trouble discriminating a baited foodwell from a nonbaited foodwell. Therefore, in terms of the performance analysis presented here, it is imperative to show that the correct orienting-response chain has in fact been elicited (0-second delay) and thus that the experimental manipulation affects only the maintenance of the response chain following an enforced delay.

Second, *it is obligatory to observe, record, and report the qualitative and quantitative nature of the animal's behavior during all phases of the trial.* Although mentioned previously, this point is emphasized because the singularly most important attribute of the delayed-response problem is that an animal's solution is virtually idiosyncratic. Because of meager experimental efforts to gain behavioral control, the performance of primates remains embarrassingly subject-unique. While these individualistic solutions must distress the experimenter who requires a specific mode of response, they represent for others the most intriguing and challenging aspect of the delayed-response problem. The ultimate control and consequent uniformity of behavior in this test demand greater specification of determining variables, together with more elaborate descriptions of individual behavior patterns correlated with these experimental manipulations.

It is with some reservation that I have championed a concerted attempt to specify the variables controlling delayed-response performance, because in so doing we may destroy the usefulness of the test. Despite its lack of control over behavior, the delayed-response problem remains one of the most sensitive behavioral tests currently available. Yet all reason would suggest that so potent a parent will sire sons equally effective as behavioral tests but more specific in their application.

References

Barnett, C. D., Ell.s, N., & Pryer, M. W. (1959). Stimulus pretraining and the delayed reaction in defectives. *Amer. J. ment. Defic.* **64,** 104.

Battig, K., Rosvold, H. E., & Mishkin, M. (1960). Comparison of the effects of frontal and caudate lesions on delayed response and alternation in monkeys. *J. comp. physiol. Psychol.* **53,** 400.

Battig, K., Rosvold, H. E., & Mishkin, M. (1962). Comparison of the effects of frontal and caudate lesions on discrimination learning in monkeys. *J. comp. physiol. Psychol.* **55,** 458.

Berkson, G. (1962). Food motivation and delayed response in gibbons. *J. comp. physiol. Psychol.* **55,** 1040.

Bliss, W. D. C. (1960). The role of perceptual cues in the delayed reaction. *J. comp. physiol. Psychol.* **53**, 176.

Blomquist, A. J. (1960). Variables influencing delayed response performance by rhesus monkeys. Doctoral dissertation, University of Wisconsin. University Microfilms, Ann Arbor, Michigan, No. 60–5717.

Braun, H. W. (1952). Effects of electroshock convulsions upon the learning performance of monkeys: II. Delayed response. *J. comp. physiol. Psychol.* **45**, 352.

Cadell, T. E. (1963). The effects of fornix section on learned and social behavior in rhesus monkeys. Doctoral dissertation, University of Wisconsin. University Microfilms, Ann Arbor, Michigan, No. 63–7586.

Campbell, R. J., & Harlow, H. F. (1945). Problem solution by monkeys following bilateral removal of the prefrontal areas: V. Spatial delayed reactions. *J. exp. Psychol.* **35**, 110.

Carpenter, C. R., & Nissen, H. W. (1934). An experimental analysis of some spatial variables in delayed reactions of chimpanzees. *Psychol. Bull.* **31**, 689.

Cowles, J. T. (1940). "Delayed response" as tested by three methods and its relation to other learning situations. *J. Psychol.* **9**, 103.

Cowles, J. T. (1941a). Food versus no food on the pre-delay trial of delayed response. *J. comp. Psychol.* **32, 153.**

Cowles, J. T. (1941b). Discrimination learning and pre-delay reinforcement in delayed response. *Psychol. Rev.* **48**, 225.

Cowles, J. T., & Nissen, H. W. (1937). Reward-expectancy in delayed responses of chimpanzees. *J. comp. Psychol.* **24**, 345.

Dean, W. H., & Davis, G. D. (1959). Behavior changes following caudate lesions in rhesus monkey. *J. Neurophysiol.* **22**, 524.

Finan, J. L. (1942). Delayed response with pre-delay reënforcement in monkeys after the removal of the frontal lobes. *Amer. J. Psychol.* **55**, 202.

Fletcher, H. J. (1964). Activity during delay interval and delayed response errors in monkeys. *Psychol. Rep.* **14**, 685.

French, G. M. (1959). Performance of squirrel monkeys on variants of delayed response. *J. comp. physiol. Psychol.* **52**, 741.

French, G. M., & Harlow, H. F. (1962). Variability of delayed-reaction performance in normal and brain-damaged rhesus monkeys. *J. Neurophysiol.* **25**, 585.

Furchtgott, E. (1963). Behavioral effects of ionizing radiations: 1955–1961. *Psychol. Bull.* **60**, 157.

Gleitman, H., Wilson, W. A., Jr., Herman, Magdalena M., & Rescorla, R. A. (1963). Massing and within-delay position as factors in delayed-response performance. *J. comp. physiol. Psychol.* **56**, 445.

Glickstein, M., Arora, H. A., & Sperry, R. W. (1963). Delayed-response performance following optic tract section, unilateral frontal lesion, and commissurotomy. *J. comp. physiol. Psychol.* **56**, 11.

Gross, C. G. (1963). Effect of deprivation on delayed response and delayed alternation performance by normal and brain operated monkeys. *J. comp. physiol. Psychol.* **56**, 48.

Harlow, H. F. (1932). Comparative behavior of primates. III. Complicated delayed reaction tests on primates. *J. comp. Psychol.* **14**, 241.

Harlow, H. F. (1951). Primate learning. *In* "Comparative Psychology" (C. P. Stone, ed.), 3rd ed., pp. 183–238. Prentice-Hall, New York.

Harlow, H. F., & Bromer, J. A. (1939). Comparative behavior of primates. VIII. The capacity of platyrrhine monkeys to solve delayed reaction tests. *J. comp. Psychol.* **28**, 299.

Harlow, H. F., & Moon, L. E. (1956). The effects of repeated doses of total-body X radiation on motivation and learning in rhesus monkeys. *J. comp. physiol. Psychol.* **49,** 60.

Harlow, H. F., Uehling, H., & Maslow, A. H. (1932). Comparative behavior of primates. I. Delayed reaction tests on primates from the lemur to the orang-outan. *J. comp. Psychol.* **13,** 313.

Harlow, H. F., Davis, R. T., Settlage, P. H., & Meyer, D. R. (1952). Analysis of frontal and posterior association syndromes in brain-damaged monkeys. *J. comp. physiol. Psychol.* **45,** 419.

Harlow, H. F., Harlow, Margaret K., Rueping, R. R., & Mason, W. A. (1960). Performance of infant rhesus monkeys on discrimination learning, delayed response, and discrimination learning set. *J. comp. physiol. Psychol.* **53,** 113.

Harrison, R., & Nissen, H. W. (1941). Spatial separation in the delayed response performance of chimpanzees. *J. comp. Psychol.* **31,** 427.

Hayes, K. J., & Thompson, R. (1953). Non-spatial delayed response to trial-unique stimuli in sophisticated chimpanzees. *J. comp. physiol. Psychol.* **46,** 498.

Heron, W. T. (1951). Learning: general introduction. *In* "Comparative Psychology" (C. P. Stone, ed.), 3rd ed. pp. 137–182. Prentice-Hall, New York.

Hunter, W. S. (1913). The delayed reaction in animals and children. *Behav. Monogr.* **2,** No. 1 (Serial No. 6).

Jacobsen, C. F. (1936). Studies of cerebral function in primates. I. The functions of the frontal association areas in monkeys. *Comp. Psychol. Monogr.* **13,** No. 3 (Whole No. 63).

McDowell, A. A. (1958). Comparisons of distractibility in irradiated and non-irradiated monkeys. *J. genet. Psychol.* **93,** 63.

McDowell, A. A., & Brown, W. L. (1958). Facilitative effects of irradiation on performance of monkeys on discrimination problems with reduced stimulus cues. *J. genet. Psychol.* **93,** 73.

McDowell, A. A., Brown, W. L., & McTee, A. C. (1960). Sex as a factor in spatial delayed-response performance by rhesus monkeys. *J. comp. physiol. Psychol.* **53,** 429.

McDowell, A. A., Brown, W. L., & McTee, A. C. (1962). Sex as a factor in delayed-response and reduced-cue discimination learning by previously irradiated monkeys. *J. genet. Psychol.* **100,** 325.

Malmo, R. B. (1942). Interference factors in delayed response in monkeys after removal of frontal lobes. *J. Neurophysiol.* **5,** 295.

Maslow, A. H., & Groshong, Elizabeth (1934). Influence of differential motivation on delayed reactions in monkeys. *J. comp. Psychol.* **18,** 75.

Meyer, D. R., & Harlow, H. F. (1952). Effects of multiple variables on delayed response performance by monkeys. *J. genet. Psychol.* **81,** 53.

Meyer, D. R., Harlow, H. F., & Settlage, P. H. (1951). A survey of delayed response performance by normal and brain-damaged monkeys. *J. comp. physiol. Psychol.* **44,** 17.

Meyers, W. J., McQuiston, M. D., & Miles, R. C. (1962). Delayed-response and learning-set performance of cats. *J. comp. physiol. Psychol.* **55,** 515.

Michels, K. M., & Brown, D. R. (1959). The delayed-response performance of raccoons. *J. comp. physiol. Psychol.* **52,** 737.

Miles, R. C. (1957). Delayed-response learning in the marmoset and the macaque. *J. comp. physiol. Psychol.* **50,** 352.

Miles, R. C., & Blomquist, A. J. (1960). Frontal lesions and behavioral deficits in monkey. *J. Neurophysiol.* **23,** 471.

Mishkin, M. (1957). Effects of small frontal lesions on delayed alternation in monkeys. *J. Neurophysiol.* **20,** 615.

Mishkin, M., & Pribram, K. H. (1956). Analysis of the effects of frontal lesions in monkey: II. Variations of delayed response. *J. comp. physiol. Psychol.* **49,** 36.

Mishkin, M., & Weiskrantz, L. (1958). Effects of delaying reward on visual-discrimination performance in monkeys with frontal lesions. *J. comp. physiol. Psychol.* **51,** 276.

Mishkin, M., Prockop, Elinor S., & Rosvold, H. E. (1962). One-trial object-discrimination learning in monkeys with frontal lesions. *J. comp. physiol. Psychol.* **55,** 178.

Nissen, H. W., & Elder, J. H. (1935). The influence of amount of incentive on delayed response performance of chimpanzees. *J. genet. Psychol.* **47,** 49.

Nissen, H. W., & Harrison, R. (1941). Visual and positional cues in the delayed responses of chimpanzees. *J. comp. Psychol.* **31,** 437.

Nissen, H. W., Carpenter, C. R., & Cowles, J. T. (1936). Stimulus- *versus* response-differentiation in delayed reactions of chimpanzees. *J. genet. Psychol.* **48,** 112.

Nissen, H. W., Reisen, A. H., & Nowlis, V. (1938). Delayed response and discrimination learning by chimpanzees. *J. comp. Psychol.* **26,** 361.

Orbach, J. (1962). Proprioceptive and positional cues in solving delayed-response problems. *Science* **135,** 667.

Orbach, J., & Fischer, Gloria J. (1959). Bilateral resections of frontal granular cortex: Factors influencing delayed response and discrimination performance in monkeys. *A.M.A. Arch. Neurol.* **1,** 78.

Polidora, V. J. & Fletcher, H. J. (1964). An analysis of the importance of S-R spatial contiguity for proficient primate discrimination performance. *J. comp. physiol. Psychol.* **57,** 224.

Pribram, K. H. (1950). Some physical and pharmacological factors affecting delayed response performance of baboons following frontal lobotomy. *J. Neurophysiol.* **13,** 373.

Pribram, K. H. (1955). Lesions of "frontal eye fields" and delayed response of baboons. *J. Neurophysiol.* **18,** 105.

Pribram, K. H. (1960). A review of theory in physiological psychology. *Annu. Rev. Psychol.* **11,** 1.

Pribram, K. H., & Mishkin, M. (1956). Analysis of the effects of frontal lesions in monkey: III. Object alternation. *J. comp. physiol. Psychol.* **49,** 41.

Riopelle, A. J. (1959). Performance of rhesus monkeys on spatial delayed response (indirect method). *J. comp. physiol. Psychol.* **52,** 746.

Riopelle, A. J., & Churukian, G. A. (1958). The effect of varying the intertrial interval in discrimination learning by normal and brain-operated monkeys. *J. comp. physiol. Psychol.* **51,** 119.

Rosvold, H. E. (1959). Physiological psychology. *Annu. Rev. Psychol.* **10,** 415.

Spaet, T., & Harlow, H. F. (1943). Problem solution by monkeys following bilateral removal of the prefrontal areas. II. Delayed reaction problems involving use of the matching-from-sample method. *J. exp. Psychol.* **32,** 424.

Spiker, C. C. (1956). Stimulus pretraining and subsequent performance in the delayed reaction experiment. *J. exp. Psychol.* **52,** 107.

Stanley, W. C., & Jaynes, J. (1949). The function of the frontal cortex. *Psychol. Rev.* **56,** 18.

Tinklepaugh, O. L. (1928) An experimental study of representative factors in monkeys. *J. comp. Psychol.* **8,** 197.

Tinklepaugh, O. L. (1932). Multiple delayed reaction with chimpanzees and monkeys. *J. comp. Psychol.* **13,** 207.

Wade, Marjorie (1947). The effect of sedatives upon delayed response in monkeys following removal of the prefrontal lobes. *J. Neurophysiol.* **10,** 57.

Wilson, W. A., Jr. (1962). Alternation in normal and frontal monkeys as a function of response and outcome of the previous trial. *J. comp. physiol. Psychol.* **55,** 701.

Wilson, W. A., Jr., Oscar, Marlene, & Gleitman, H. (1963). The effect of frontal lesions in monkeys upon widely-spaced delayed-response trials. *J. comp. physiol. Psychol.* **56,** 237.

Yerkes, R. M., & Yerkes, D. N. (1928). Concerning memory in the chimpanzee. *J. comp. Psychol.* **8,** 237.

Chapter 5
Associative Problems

Gilbert M. French

Department of Psychology, University of California, Berkeley, California

I. INTRODUCTION

An associative problem exists whenever an animal must choose among alternative responses or directions of response to produce some distinctive subsequent event, or outcome. Typically only one alternative yields an outcome which an animal would work to acquire or avoid. As a basis for successful choice, an animal must utilize its experience in the same or similar situations. To describe the variables influencing response selection has been the goal of students of problem-solving, who are interested in how differential responses in the presence of recurrent environmental regularities may be learned, retained, and transferred.

Over the years, discussions of the genre of research treated in this chapter have appeared under headings such as "symbolic processes," "higher mental processes," "cognitive processes," and "complex processes." In alluding to mechanisms which are at best dimly appreciated, these terms seem awkward and pretentious. The term "complex problems" is better, but its use is accompanied by a nagging suspicion that the major part of complexity resides in our present inability to identify readily a number of fundamentally simple sources of problem difficulty. The term "associative problems" is less objectionable and more happily descriptive. The problems discussed here are related to instrumental conditioning. There is at least an operational continuity between the procedures of associative conditioning tasks and associative problem-solving tasks. In both, experimentally-arranged correlations between events of significance and relevance for subjects are the source of systematic performance.

During the present century, researchers concerned with associative problem-solving have turned for their data from the anecdote to the experiment and from the field to the laboratory. Using artificial and increasingly simplified environments, they have been able to dissect and analyze the solution of problems and to arrive at concepts that give promise of being generally applicable.

A large and very fruitful part of the experimental work has been done on nonhuman primates. The abilities of these animals are of some intrinsic interest. Since they are man's nearest relatives in the animal kingdom, we look at them to understand better the origins of our own behavior. Because their brains are more like ours than are those of other animals, they are uniquely valuable in the study of certain neuropsychological correlations. But even beyond these anthropocentric advantages, nonhuman primates have been used for research in problem-solving for the compelling reason that they make admirable subjects. They are often tractable, highly dependable, and quick to react. When appetite alone cannot keep them working in the laboratory, their storied curiosity may suffice. Their intellectual development is slow enough to allow ample time for longitudinal research but fast enough to prevent tedium. Techniques for maintenance and testing have been very well worked out for a number of kinds of monkeys and apes. Availability of such methods is not the least part of an animal's attractiveness as an experimental subject.

The main purpose of this chapter is to depict the present status of research with nonhuman primates on several problem-solving tasks. The approach I have taken is to develop a taxonomy of methods, to demonstrate how different problems may be applied, to describe some of the results that have been obtained, and to delineate some of the relations among problem types. On occasion, in the course of advancing an argument, I have drawn upon research conducted with other animal forms. This is particularly true in instances of gaps in the primate literature.

One major kind of problem, delayed response, is the subject of Chapter 4 by Fletcher. Studies of probability learning (see Chapter 7 by Warren) and of concept formation are not treated specially here. In general, they involve elaborations of the techniques that are discussed.

In the study of associative problems, as in other areas of psychology, the invention of a test instrument has often antedated by years or even decades the discovery of variables that profoundly influence the results obtained. Meanwhile, premature and spurious interpretations have been difficult or impossible to avoid. The testing of tests continues unabated in the 1960s. Although psychology is perhaps better equipped than ever before to devlop a comprehensive theory of associative problem-solving, there should be no illusions about the completeness of our knowledge. Each issue of the *Journal of Comparative and Physiological Psychology* or the *Journal of Experimental Psychology* contains enough new mystery to dispel complacency. Even those workers dealing exclusively with discrimination learning are at no loss to find formidable theoretical difficulties. Thus, the present chapter is an interim statement rather than a final report.

II. DISCRIMINATION PROBLEMS

A. Simultaneous Discrimination

In this commonly-used method, the subject must choose between two or more possible loci of reward. In the Wisconsin General Test Apparatus (WGTA; see Chapter 1 by Meyer *et al.*), for example, a cube and a pyramid may be presented together on a test tray. The objects are both manipulanda and discriminanda. Insofar as they are merely things to be moved to uncover the subjacent foodwells, they are manipulanda. Insofar as they differ in form and in relation to distinctive outcomes, they are discriminanda. On every trial the pyramid is contiguous with reward and the cube with nonreward. Since these discriminandum-outcome relations prevail in spite of irregular trial-to-trial variations in the positions of the objects on the tray, the problem is one of a class of *nonspatial discrimination* problems. By way of contrast, there are *spatial discrimination* problems, in which the position of a given outcome is fixed. For instance, reward might always be available in the left foodwell, never in the right. The spatial discriminanda are all those stimuli that enable the subject to respond selectively to the two sides of the test tray.

Even though *two-choice* problems are more familiar, *multiple-choice* problems with several negative discriminanda are sometimes encountered. The discriminanda may be nonspatial (Klüver, 1933) or spatial (Spence, 1939). Spatial discriminanda necessarily differ, but nonspatial discriminanda may be identical. Multiple-choice problems with identical negative discriminanda are also known as "one-odd oddity problems" (see Section V, C).

During a series of researches on nonspatial discrimination by chimpanzees (*Pan*), Nissen and McCulloch (1937a) presented single positive discriminanda in combination with either 1 or 9 identical negative discriminanda. Subjects could continue to respond on each trial until they had obtained reward. Multiple-choice problems were solved in fewer total trials and with fewer trials in which there was at least one error. However, errors were less frequent on two-choice problems. When only one response was allowed on each trial, multiple-choice problems with 11 negative discriminanda were solved in fewer trials and with fewer errors than were two-choice problems (McCulloch & Nissen, 1937). Nissen and McCulloch (1937b) found evidence for the development of consistent responding to new odd discriminanda on their first appearance in multiple-choice color-discrimination problems. This capability was not readily transferable to pattern-discrimination problems.

In most studies involving simultaneous discrimination, a given set of discriminanda reappears on each trial until some criterion of performance has been reached or until some limited number of trials has been completed. In *concurrent* discrimination (sometimes called serial discrimination), the members of a list of several sets of discriminanda are presented on successive trials. On each run through the list, any single set is displayed once and only once.

Hayes *et al.* (1953) gave chimpanzees concurrent object discriminations in lists of 1, 5, 10, and 20 pairs of discriminanda. For equal numbers of runs, accuracy of choice was inversely proportional to the length of the list, a relation which held for every one of the six subjects. Hayes *et al.* noted parenthetically that an occasional change in the order of presenting the pairs had no adverse effect on performance.

Leary (1957) systematically studied the consequences of rearrangement of paired objects on successive runs in concurrent discrimination. Rhesus monkeys (*Macaca mulatta*) were given seven runs through each of three 9-pair lists. For the control list (condition C), identical pairs appeared in the same serial order on each run. For a second list (condition P), the pairs remained unchanged, but were presented in a varied serial order on every run. For a third list (condition O), the order of positive objects was the same on each run, but the order of the negative discriminanda was changed so that they were paired with different positive objects on successive runs. During the seven runs, the separate pairs in condition P and the separate negative objects in condition O were presented no more than once at any given ordinal position within the list. Under all conditions, performance improved from near-chance levels on run 1 to very high levels on run 7. The lack of significant differences between conditions C and P indicated that the shuffling of unchanged pairs is without effect, confirming in a rigorous way the incidental observations of Hayes *et al.* However, when the pairs themselves were reconstituted between runs in condition O, performance lagged behind that of the other conditions on runs 5 to 7. In explaining the pattern of impairment under condition O, Leary emphasized the principles of discrimination learning that had been developed by Spence (1936, 1952). By assuming independent changes in the excitatory strength, or "attractiveness," of positive and negative discriminanda, Leary was able to show how the results might have been predicted. On earlier runs the negative objects had not been responded to as consistently under condition O as under the conditions in which pairs had been retained from one run to another. As a result of fewer nonrewarded responses, the negative objects remained more attractive on later runs under condition O than under the other conditions.

B. Successive Discrimination

The second fundamental method of training is that of successive discrimination, also called the *differentiation*, or *go–no-go* method. The paradigm is somewhat similar to that of differentiation in classical conditioning. One manipulandum and one discriminandum appear on each trial. Over a series of trials, different discriminanda are associated with different outcomes. Since the subject has learned to use the manipulandum to produce reward before the start of formal successive training, it must learn to inhibit the response in the presence of the negative discriminandum. The inhibition may be shown either by an increase of reaction time or by the complete lack of the response in a trial of fixed duration. Errors in successive problems have customarily been defined as nonresponses to positive discriminanda and as responses to negative discriminanda. The latter kind of error is far more prevalent than the former, even at the beginning of training. Despite the difference in performance criteria for the two kinds of error, the successive method is particularly useful whenever there are liable to be interactions among discriminanda, as could be expected with two odors or two tones.

In its pure form, successive discrimination has been investigated seldom in primates. The only extensive analytical study of the method is that of McClearn (1957) on object differentiation by cynomolgus monkeys (*Macaca irus*). McClearn's report is confined to intraproblem effects occurring in a long series of 6-trial problems. McClearn began by training his experimentally naive monkeys to move a gray wooden cube to obtain food reward. He then gave pretraining designed to adapt the subjects to displacing objects that differed from one another in many stimulus characteristics (e.g., size, form, color, and texture). Thirty different objects were presented for six consecutive trials each, and a response was rewarded on every trial. At this time, of course, the objects functioned only as manipulanda. During this pretraining, means and variabilities of response latencies decreased as performance became more stable. Successive-discrimination training consisted of 432 six-trial problems in which both positive and negative objects appeared. A given object was presented for two, three, or four trials in one of 12 different 6-trial sequences. On the first presentation of either a positive or a negative discriminandum, there were fewer than 5% nonresponses. With later appearances of the positive object, the percentage of nonresponses remained low. On the second appearance of a negative object, however, the percentage of nonresponses jumped to between 40 and 50, and the mean latency of responses increased reliably. Additional presentations of a negative object did nothing to accentuate these effects. It was also found that when "a rewarded stimulus is followed by an unrewarded one, there is no effect upon the tendency to respond to the rewarded stimulus on its next presentation, but an unrewarded stimulus followed by a rewarded one increases the response tendency to the unrewarded one on its next presentation" (McClearn, 1957, p. 440). This result, obtained for both nonresponse and latency measures, *may* account for the failure of response inhibition to increase beyond the second no-go trial, but an exact determination remains to be made.

Are successive spatial discriminations possible? Because spatial discriminanda usually do not vary over trials, one might at first think not. The common practice of using a centered manipulandum of unchangeable position heightens this impression. Yet, a stationary manipulandum is not required by the differentiation method. Over a number of trials, for example, a single food-container could be moved at random from one side of the apparatus to the other. A response could be rewarded if the container appeared at the right and not rewarded if it appeared at the left. Since only positional cues are given, the problem qualifies as a successive spatial discrimination. To the spatial discriminanda that remain the same from trial to trial, this procedure would add certain trial-specific spatial discriminanda. These are the different positions of the manipulandum in the test apparatus. This precise form of successive discrimination seems not to have been given to experimental subjects.

Mishkin and Pribram (1956, Exp. 2) presented a very closely related problem to rhesus monkeys. In their procedure, the response to a centrally-located food-container was rewarded if an object was shown at the left and not rewarded if the same object was shown at the right. There was a delay between the appearance of the object and the subject's opportunity to respond, and the problem was very difficult. No monkey reached the solution within the 500-trial limit of training.

Simultaneous and successive discrimination differ in at least two ways. One difference is in the opportunities they afford for comparison of stimuli. In simultaneous problems, the structure of the experimental apparatus often permits an animal to view the different discriminanda in close temporal proximity. In successive problems, some minimal delay between presentations of discriminanda is part of the method. Riley *et al.* (1960) have demonstrated that rats learn a simultaneous brightness discrimination more slowly when a barrier between the discriminanda interferes with comparison. The impairment produced by the barrier is greater for problems in which the difference in intensity between the two discriminanda is smaller. Any interval of time between observations of separate discriminanda probably diminishes effective comparison. To the extent that adequate performance depends on comparison, the delay built into the successive method should be detrimental.

Another difference between the simultaneous and successive methods is in the nature of the required responses. In simultaneous problems, reward is at least potentially available on all trials. The subject must learn only to direct its response appropriately. In successive problems, no reward can be obtained on some trials, and the acquisition criterion demands that the animal learn not to respond to the manipulandum on these no-go trials. This kind of learning is most likely the development of some new response to take the place of the old. The new response, unspecified and uncontrolled by the experimenter, is made somewhere away from the manipulandum and competes with the response to the manipulandum. French (1959) has described a number of techniques devised by individual squirrel monkeys (*Saimiri sciureus*) to prevent responding to the manipulandum when the negative discriminandum has been presented. The orientation of these animals in the test apparatus suggests that they shun even the sight of the manipulandum on no-go trials. For a time, at least, the new response may be much more effortful than the one it replaces. As in simultaneous problems, the direction of response is important, but in successive problems more than direction is involved. The response required on a no-go trial differs markedly in structure from those required on go trials.

In a searching and valuable methodological discussion, Weiskrantz (1957) has suggested a technique that may well both facilitate acquisition and reduce the usually great difference in frequency between errors of omission and errors of commission. The procedure is to reward nonresponses on no-go trials. At the end of a trial in which the response has been withheld successfully, food is delivered at some distance from the manipulandum, for instance in a shallow cup attached to the end of a metal rod (Gross, 1963a; Gross & Weiskrantz, 1962). This method appears practicable and successful.

C. Conditional Discrimination

The simultaneous and successive paradigms are used to investigate performance with single sets, usually pairs, of discriminanda. Conditional discrimination allows the study of multiple, intersecting sets, each of which conveys information relevant to the solution of a problem. In the simplest of conditional discriminations, there are two separable sets of discriminanda: one simultaneous and one successive. All members of the simultaneous set are present on all trials. Only one member of the successive set appears on any trial. It signals the positive and negative members of the simultaneous set for that trial.

One kind of conditional problem is *sign-differentiated nonspatial discrimination,* in which the simultaneous discriminanda are nonpositional. Imagine an object discrimination involving a chalkboard eraser and a funnel. When the two objects appear on a red test tray, the eraser covers food, and when they appear on a yellow tray, the funnel covers food. The simultaneous discriminanda here are the objects, and the successive discriminanda are the tray colors. Spence (1952) prefers to speak of stimulus compounds. There would be two positive compounds, "eraser-red" and "funnel-yellow," and two negative compounds, "funnel-red" and "eraser-yellow."

A study of sign-differentiated nonspatial discrimination by children (Gollin & Liss, 1962) illustrates some of the variables to be considered in designing conditional-discrimination problems. The simultaneous discriminanda were white triangle and white circle, the forms appearing against either a black background or a black-and-white-striped background. The children were trained first with cards having one kind of background, either black or striped. Then they were trained on reversed discriminandum-outcome relations with cards of the other background, next on a simple trial-to-trial alternation of backgrounds, and finally on a random temporal pattern of alternation. At each stage the criterion of performance was 10 consecutive trials without error. Under the first two conditions the discrimination of white triangle and circle was easier on the black background than on the striped background. This finding was similar to that obtained with rats by North *et al.* (1958). The result indicated to Gollin and Liss that "variations in background may introduce certain perceptual difficulties which interfere with the establishment of either the initial or subsequent discriminations independent of the configural or compound features with which such experiments are concerned" (Gollin & Liss, 1962, p. 852). Only among the youngest of the three groups of subjects, children between 3.5 and 4 years old, was there reliable negative transfer from the original to the reversal task. Following reversal, however, the change to alternation of background stimuli dramatically disrupted performance in all age groups, the effect being most pronounced in the youngest children. Apparently the isolated training with each background discriminandum was not very readily transferable to a situation involving alternation of the same discriminanda. Even in the oldest group, between 5.5 and 6 years old, the criterion was attained only after a mean of 24.3 trials and 6.7 errors. The children who reached criterion on the simple alternation were able to shift to a random change of background without added difficulty.

Warren (1964) showed that sign-differentiated nonspatial discriminations can be learned quickly even though the successive discriminanda are changed frequently from the very start of training. The simultaneous discriminanda were thick wooden plaques differing in regular geometrical outline. The successive discriminanda were of three types: color of plaque (black or white), plane of presentation of plaque (horizontal or vertical), and a combination of color and orientation. Under each condition, experimentally sophisticated rhesus monkeys met the criterion of 10 consecutive correct responses with a mean of fewer than five errors. Learning was reliably more rapid when the two kinds of successive discriminandum were combined than when either was used singly. Warren's subjects had been well adapted to the WGTA used in this experiment and had been trained on simultaneous discriminations and on other kinds of conditional discriminations, so they probably had overcome most of the major impediments to fast learning before this experiment began.

In the sign-differentiated nonspatial problems discussed so far, the successive discriminanda have been added stimuli, such as color of ground, color of figure, and orientation of figure. Problems can be designed in which the successive cues are provided by relations among members of the set of stimultaneous discriminanda. These are called *ambivalent-cue* problems. Consider three objects that differ from one another in color: red (R), green (G), and blue (B). Depending on the particular objects that are paired, a given one may be either positive or negative. Discriminandum-outcome relations for the three unique pairings might be R+ versus G—, B+ versus R—, and G+ versus B—. As they are used here, the color cues are ambivalent in that each is equally often associated with reward and nonreward. To solve problems in which pairings are changed irregularly from trial to trial, the subject must observe both discriminanda before responding.

There is no well-developed theory of how to teach animals to respond successfully in ambivalent-cue problems. The practice usually followed has been to train them on separate configurations of the discriminanda first and then to combine several pairings in the manner of a concurrent discrimination problem. Among primate subjects, this method has been used with chimpanzees (Nissen, 1942, 1951) and with macaque monkeys (Noer & Harlow, 1946). The method deals with the development of appropriate observing responses only indirectly. Its operational emphasis is on the subject's learning to perform with frequent shifts among possible pairs of discriminanda.

In *sign-differentiated position discriminations*, the simultaneous discriminanda are spatial. Successive discriminanda may take a variety of forms. Harlow (1942) used the presence and absence of food under a "sample" object as successive discriminanda in a series of experiments on rhesus monkeys. For this research, a test tray was divided into two unequal areas by a low strip of wood. In the narrower area at the experimenter's left, a single object appeared. In the wider area at the right, two identical objects were presented side by side. Subjects were pretrained always to respond to the isolated object first. If the first response were rewarded, the object at the right of the remaining pair would be positive. If the first response were unrewarded, the object at the left of the remaining pair would be positive. After this problem was solved, it was presented concurrently with other problems, as described in Section V, E.

Pribram and Mishkin (1955, Exp. 2) trained two normal and eight brain-damaged rhesus monkeys on a sign-differentiated position discrimination in which the successive discriminanda were objects, tobacco tin and ash tray. On each trial an object was exposed at the center of a display panel between identical food-containers that were mounted to the right and left. No contact with the objects was demanded by the procedure. For half the subjects, tobacco tin was correlated with food in the left container and ash tray with food in the right container. For the rest, these relations were reversed. The problem was solved by all monkeys but proved very difficult. Probably an important source of difficulty was the distance between the centrally placed object and the laterally placed food-containers, with which the only manual contact took place. As shown by many of the very recent studies reviewed in Chapter 1 by Meyer *et al.*, separations of cue and response in discrimination problems can impair performance seriously.

That the spatial separation indeed contributed to the relative difficulty of the conditional discrimination was indicated in later research on eight of the same monkeys (Mishkin & Pribram, 1956, Exp. 2). The conditional problem reappeared in combination with a delayed-response condition. Under this procedure, no animal gave any sign of learning during 200 trials. The procedure was then revised so that on every trial one of the successive discriminanda appeared over the same food-container. For some animals this was always the container at the right; for others it was always the container at the left. With the change of training technique, seven of the eight subjects reached a criterion of 90 correct in 100 trials during the 500 trials of additional training on the problem.

III. DISCRIMINATION REVERSAL PROBLEMS

Reversal problems involve a change, or shift, in the relation between members of an invariant set of discriminanda and particular outcomes. Positive discriminanda before reversal become negative thereafter, and negative become positive. Reversal problems can be presented in either simultaneous or successive settings.

Results of classic research on transfer of training usually showed that interference occurs when new responses are required to old stimuli. In discrimination-reversal experiments the effect lacks stability, particularly if subjects are given experience in making reversals. By giving specific practice on reversals, experimenters have not only diminished negative transfer but also enabled their subjects to approach maximally-efficient performance quite closely. In a wide variety of studies (see Chapter 7 by Warren), subjects have been trained on a series of reversals using a single set of discriminanda. For example, the cumulative effect of repeated reversals of a positional right-left discrimination has been investigated with rats by Dufort *et al.* (1954) and with human mental defectives by House and Zeaman (1959). Trained to a criterion before each shift, either kind of subject made more errors on the first few reversals than on the initial discrimination. But, within six shifts, mean errors per reversal dropped to almost one and remained at that level. Since the only source of cues to reversal was the change of relations between discriminanda and outcomes, a single error per reversal was a limiting value.

Somewhat similar improvement has been shown on repeated object-discrimination reversals by chimpanzees (Schusterman, 1962) and by rhesus monkeys (Gross, 1963b). Schusterman required criterional performance before reversal, and Gross trained his animals for a predetermined number of trials before reversal. In both studies, the approach to optimal performance was slower than for either the rats of Dufort *et al.* or the retardates of House and Zeaman. This difference may eventually be traceable to the nature of the discriminanda, to the apparatus, to the species, or to some interaction of these variables.

Schusterman's research was of special value in demonstrating that strategies acquired during multiple discrimination reversals with a single pair of objects are transferable to other, related problems. Performance both on two new repeated-reversal problems and then on object-quality-discrimination-learning set was markedly facilitated (see Chapter 2 by Miles).

Instead of training subjects on repeated reversals with one pair of discriminanda, an experimenter can give a series of reversal problems with different pairs, each of which is retained for but one reversal (see

Chapter 2 by Miles). The development of nonspatial-discrimination-reversal-learning sets in this manner has been shown for rhesus monkeys by Harlow (1950b) and Meyer (1951), for squirrel monkeys by Rumbaugh and McQueeney (1963), and for marmosets (*Callithrix*) by Cotterman *et al.* (1956). From one problem to another, the reversal begins after varying numbers of trials in order that no cue be provided by the ordinal position of the first reversal trial within each problem. Following several reversals, a subject's intraproblem performance rises from a much-less-than-chance level on the first reversal trial to a much-greater-than-chance level on the second and later reversal trials. The results indicate again that a reversal of the direction of response can be acquired by animals that have had but one exposure to appropriate stimulation.

The study by Harlow (1950b) showed that the introduction of reversals when animals had previously been trained only on discrimination problems temporarily interfered with performance in an unexpected but unambiguous way. Trained by Harlow (1950a) on 232 object-discrimination problems, seven rhesus monkeys and a mangabey (*Cercocebus*) were given additional tuition on 112 object-discrimination-reversal problems. During the latter problems, errors on second reversal trials gradually decreased from 39% in the first 14 problems to 2% in the last 14 problems. A portion of the reversal errors could be traced to a response tendency which had been eliminated in the earlier discrimination-learning-set training and which remained inactive on prereversal trials. These "differential-cue errors" (see Chapter 2 by Miles) appeared on those postreversal trials in which the relevant nonspatial discriminanda were first changed in their positions on the test tray, indicating a temporary and situation-specific reversion to the use of irrelevant positional cues.

Riopelle and Copelan (1954) showed how a stimulus change other than that of discriminandum-outcome relations could be used as a cue to reversal (see Chapter 2 by Miles). They first trained rhesus monkeys on a series of object-discrimination reversals in which the color of the test tray was changed from green on prereversal trials to yellow on postreversal trials. At the end of the initial phase of the experiment the animals had learned to use the specific change of tray color as a cue to correct response on the first reversal trial. After further training on a series of other color changes, the monkeys could make errorless reversals on their initial exposures to new changes of color. The reversal cue had become generalized.

Yet another kind of reversal cue can be provided by a change in the spatial arrangement of sets of simultaneously-presented discriminanda. Crawford (1962) used a framework that permitted objects to be displayed along horizontal, vertical, or diagonal lines in the frontal plane of the subject. At the beginning of reversal, the line of array of discriminanda was changed. There were 288 reversal problems, 48 each of the six possible sequences of two different lines of array. Spider monkeys (*Ateles geoffroyi*) and cynomolgus monkeys learned to use the change in line of display as a cue to immediate reversal of response. Within the limits of training, however, cebus monkeys (*Cebus albifrons*) gave no evidence of improved performance on first reversal trials, even though they came to excel the cynomolgus monkeys on second reversal trials. Crawford suggested that the cebus monkeys may need the experience of changed discriminandum-outcome relations that is provided by the first reversal trial.

The outcome of the first reversal trial, color of test tray, and arrangement of stimuli are all discriminanda in their own right. Their function is like that of the successive discriminanda in conditional discrimination problems. As this conception is grasped, the essential similarity of discrimination reversal and conditional discrimination becomes manifest. In both kinds of problem, the relations between simultaneous discriminanda and outcomes fluctuate over trials. In both, moreover, these changes are heralded by another class of discriminandum. The transition between discrimination reversal and conditional discrimination is to some extent vague and arbitrary. Distinctions between the two classes of problem are based mainly upon the relative stability of given discriminandum-outcome relations from trial to trial.

But one should also recognize that it has been traditional in discrimination-reversal problems to use as successive discriminanda only the relations of the simultaneous discriminanda with particular outcomes, and to allow no cue to the shift immediately before the responses of record on reversal trials. With these procedures of classic discrimination reversal, the subject cannot anticipate the change on the first reversal trial. Neither can it later confirm from some external source that a change has occurred on some previous trial. There is no prompting, no mnemonic device by courtesy of the experimenter. In conditional problems, on the other hand, some form of prompting is essential to efficient performance when the correlation between discriminanda and outcomes changes frequently and irregularly.

IV. ALTERNATION PROBLEMS

A. Single Alternation

All problems of this type are characterized by the regular alternation of discriminandum-outcome relations from trial to trial. In *spatial alterna-*

tion the sequence of rewarded positions, right (R) and left (L), might be either RLRL . . . or LRLR . . . on consecutive trials. It is usual on the first trial either to force the response to the baited locus or to reward the animal's preferred response. After that, of course, the animal must alternate its direction of response in order to obtain rewards consistently. The procedure in *nonspatial alternation* is analogous, the only difference being that the subject must alternate between discriminanda the positions of which are varied. Under the trial-rerun correction method often used in alternation experiments, the subject must continue to respond with the reward unshifted after making an error. As in all applications of the correction method, the definition of a "trial" becomes inconsistent. No longer does a trial necessarily contain only one response. It may contain any number, all but one of which are errors. The method has two characteristic and possibly advantageous features. One is that each trial terminates in a correct response, after which alternation is resumed. The other is that a simple strategy of "shift" of response is adequate for solution; during learning, the noncorrection method requires a "win-shift, lose-stay" strategy in which the correct response is conditional on the *outcome* of the previous response. Whether correction is superior to noncorrection in alternation training has not yet been determined.

Wilson (1962) attempted to answer the question by training rhesus monkeys concurrently on spatial alternation under both procedures. In presolution performance there was no significant difference between conditions, but no definitive test was provided as the monkeys never greatly exceeded chance success on the problem.

A particularly interesting aspect of Wilson's experiment was the comparison of contingent and noncontingent techniques of informing subjects about discriminandum-outcome relationships. In most two-choice problems, the procedures limit this information to whatever outcome is produced by the response to the manipulandum. With all response-contingent procedures, subjects have no information about the outcome that would have resulted from the alternative response. With noncontingent procedures, however, a response to one manipulandum not only produces the usual outcome at that locus but also displays the outcome at the other locus (e.g., Bush & Mosteller, 1955; Wilson & Rollin, 1959). Under appropriate conditions, which seem not to have been well defined, one might expect facilitation of learning when a noncontingent procedure is followed. It appears that a minimum requirement is getting the animal to observe the "free" information. The monkeys in Wilson's study derived no relative benefit from the added noncontingent information, although it is moot whether the lack of difference between conditions represented failure to receive stimulation or failure to act on it. At least the issue has been brought forward for investigation.

Procedures in most alternation experiments require a relatively small intertrial period for reconstituting the stimulus display. When an experimenter extends this interval to any desired length, a *delayed alternation* problem results. The method is markedly similar to the delayed-response method described in Chapter 4 by Fletcher. In both problems, cues and responses are separated in time. Regardless of the specific sources of cues to the response on trial n in delayed alternation, they could not have originated any later than the response phase of trial $n - 1$. The intertrial interval of delayed alternation is thus comparable to the delay phase of delayed response.

Similar results are sometimes but not always obtained with delayed response and delayed alternation. Gross (1963c) found that an increase in food deprivation facilitated direct-method spatial delayed response but had no effect on spatial delayed alternation by rhesus monkeys working for raisins. The contrast is attributed by Gross to the greater stimulus value of the raisins shown as cues during the baiting phase of the delayed-response task. This hypothesis is congruent with the concept that the "bait serves as a cue in delayed response, but not in delayed alternation" (Gross, 1963c, p. 48).

Authors of textbooks in psychology have usually distinguished between delayed response and delayed alternation in terms of the sources of cues to solution (e.g., Morgan, 1961, p. 291; Osgood, 1953, p. 657). It is agreed that in delayed response the critical cue is given by the experimenter on each trial. It consists of the display of food itself (in the direct method) or the presentation of some stimulus that has been paired repeatedly with food (in the indirect method). In any event, the cue originates from outside the animal. Now, in delayed alternation, so the analysis goes, the experimenter does not directly provide a cue. The cue comes from inside the animal and is believed to be "primarily motor" (Osgood's phrase) in the sense that turning in one direction on trial n is the signal to turn in the opposite direction on trial $n + 1$. The theory is seriously inadequate because motoric or somesthetic cues are far from necessary for the performance of delayed alternation. Rats can carry out spatial delayed alternation even though they have been anesthetized during the intertrial interval (Ladieu, 1944; Loucks, 1931). Rhesus monkeys are capable of nonspatial delayed alternation in which any stereotyped method of reorienting responses in space would be useless (Behar, 1961; Pribram & Mishkin, 1956). Although the motor theory may yet account for results obtained in some settings or at some stages of learning, it is unacceptable as a general explanation of the cuing of delayed alternation.

The fact is that the sources of cues to successful delayed alternation are today neither intuitively obvious nor experimentally demonstrated.

One can properly speak only of possibilities. Since the noncorrection method of alternation training requires subjects to use outcomes of responses as conditional cues, the proposition that outcomes can determine choices effectively in this kind of problem is worthy of appraisal. Let us think of food reward. Even should food be a cue to alternation, the information would have to be used quite differently from the way it is used in delayed response and discrimination. To solve the latter problems, the subject must *approach* the locus at which food was last perceived. To solve the alternation problem, it must *avoid* that locus. Not only must the animal shift away from the location of previous reward, but it must do so in the absence of any special experimenter-provided cues to reversal. In the forms of discrimination reversal discussed earlier in this chapter, a shift after one nonrewarded trial is the best that can be expected when such cues are missing. Alternation demands zero-trial reversal.

An experimental demonstration by Riopelle and Francisco (1955) is germane to the present argument. The investigators gave three rhesus monkeys 250 six-trial discrimination-learning problems in which the first-trial response was always rewarded because both objects covered food. The object not chosen on trial 1 was then correct on trials 2 through 6. Performance on trial 2 remained at the same low level over the entire course of training, but accuracy on trials 3 through 6 improved steadily. The animals were unable to reverse the direction of their response on the basis of the first trial alone. Obviously the data gave no support to any notion that reward can serve as a cue to reversal, as required by delayed alternation. Could nonreward serve as a cue to repetition of a response? Riopelle and Francisco assigned another group of monkeys to a condition in which the first-trial response was never rewarded. The same object that had been chosen on trial 1 was, however, correct on trials 2 to 6. Under this condition, performance on trial 2 improved progressively over the 250 problems. The monkeys could learn to repeat a nonrewarded response, even though they had no strong inclination to do so at the beginning of training. This suggests a possible mechanism for self-correction in delayed alternation, but it cannot explain consistent correct responding.

Consider for a moment the procedures of discrimination reversal and of the problems used by Riopelle and Francisco. In both, the same discriminandum-outcome relations persist over most trials, and the mode of solution that usually works is one of "win-stay, lose-shift." As noted by Behar (1961), this strategy is diametrically opposed to the win-shift, lose-stay strategy demanded by noncorrection alternation problems. To evaluate most effectively the possibility that animals can use outcome cues in delayed alternation, one should look for procedures that maximize the likelihood of a subject's adopting an appropriate strategy and minimize the likelihood of its adopting a conflicting strategy. By modifying 2-trial discrimination problems, exactly the right conditions for this appraisal can be produced. A uniform shift or win-shift strategy is demanded if the response on trial 1 is always rewarded and if the other discriminandum always marks the locus of reward on trial 2. A uniform stay or lose-stay strategy is demanded if the response on trial 1 is never rewarded but the discriminandum that was chosen is always associated with reward on the next trial. The relevant experiments have already been done (Brown & McDowell, 1963; McDowell & Brown, 1963a, 1963b; Mishkin *et al.*, 1962). The results are clear and amply document that rhesus monkeys can learn to adopt either kind of strategy (see Chapter 2 by Miles).

It appears possible, then, in 2-trial discrimination problems to cue a change in direction of response on trial $n+1$ by presenting a reward on trial n. As in other kinds of associative problems, outcomes could have a cuing function. The main requirement seems to be a degree of consistency or predictability in the way in which outcomes are related to specific discriminanda within problems. The results should restrain out-of-hand rejections of the concept that experimenter-provided outcome cues may be used in delayed alternation. Whether such cues *are* used is another question and one for which only a partial answer is available.

The data Wilson (1962) obtained under his noncorrection condition permitted the analysis of how alternation can depend both upon the outcome and upon the response of the immediately preceding trial. Among rhesus monkeys that had not yet attained stable performance, percentages of alternating were greater after an alternation than after no alternation. From this result, sequences of alternation and of perseveration to the same side may be inferred. Alternation was more frequent following nonreward than following reward. In the early stages of alternation learning, then, lose-shift behavior has some prominence. Schusterman and Bernstein (1962) had similar results in an attempt to train gibbons (*Hylobates lar*) on spatial alternation without correction. The apes abandoned their initial positional preferences and took up response alternation before giving any sign of mastering the problem. During the last third of training, the high incidence of consecutive errors showed that the subjects were often alternating away from the positions of nonreward. The appropriate response-alternating tendency thus existed before it was applied in a way sufficient to solve the problem. This occurred both in a group trained only on alternation and in a group that was trained concurrently on alternation and object discrimination. The latter group maintained nearly perfect discrimination performance at the same time that alternation performance was at a chance level.

An experimental analysis by Behar (1961) of response tendencies in nonspatial delayed-alternation learning identifies potential sources of difficulty other than those directly attributable to discrimination and memory. Six rhesus monkeys, which had previously formed object-discrimination-learning sets, were given 48 trials per day for 40 days on object-alternation tasks. For three animals the training was divided into blocks of six trials, each with a unique pair of discriminanda; for the other three animals the division was into blocks of 24 trials. Both groups improved significantly with practice, and there was no reliable difference in learning rates or over-all performance. Even at the end of training, after 1,920 trials, the mean percentage of correct responses for both groups combined was less than 70. No strong perseveration to position was seen, and it may be presumed that positional preferences had been reduced to a rather low level by earlier object-discrimination training. Originally, the animals tended to perseverate in responding to an object that had covered food. The frequency of such responses diminished rapidly during the first half of training and remained slight thereafter. Continued responding to an object chosen on the first trial with a new pair was at first infrequent, gained rapidly in importance, and then diminished as training progressed. Such object preferences appear to have been eliminated faster in the group that had only six trials with each pair of objects. Possibly a major source of difficulty in the learning of alternation was the near-zero initial probability of the tendency to alternate. Thus, the animals may have had trouble "in discovering the correct response" (Behar, 1961, p. 542).

Go–no-go single alternation, invented by Mishkin and Pribram (1955), demands alternation of responding and not responding to a single manipulandum. Mishkin and Pribram placed a small box with a movable lid in the center of the WGTA test tray. Before each "go" trial, the box was baited with a peanut during the 5-second interval in which the opaque screen of the apparatus was lowered. The first response of a session was always rewarded. After that, the box contained no peanut on every other trial. Subjects were required to withhold a response to the manipulandum for 5 seconds on "no-go" trials. The only plausible sources of differential cues in this problem were the animals' previous responses and the outcomes of previous trials. Mishkin and Pribram found that rhesus monkeys could reach very high levels of performance within 500 trials. Even animals with frontal-lobe lesions were capable of learning go–no-go alternation with relative ease.

Procedural parallels between the more usual double-manipulandum forms of delayed response and delayed alternation readily suggest a delayed-response analog of go–no-go alternation. Mishkin and Pribram (1956) administered a *go–no-go delayed-response* task to the same monkeys that had been trained previously on go–no-go alternation. The single, centered food-box was used again. At the beginning of a trial under the direct method of testing, a peanut or an empty hand was shown above the box. When the bait, if any, had been placed in the box, the lid was closed and a 5-second delay started. At the end of the delay phase, the test tray was moved to within reach of the subject, where it remained until the response or until 5 seconds had elapsed without a response. Trial-rerun correction was employed. Within 250 trials, all subjects reached a criterion of 90 correct responses in 100 consecutive trials. Again, frontal animals had no particular difficulty in reaching a sustained high level of performance.

If the successive forms of delayed alternation and delayed response present any major obstacle to efficient performance by sophisticated rhesus monkeys, it stems from the requirement of response inhibition on no-go trials. The high frequency of repetitive errors on such trials at the start of training (Pribram, 1959, Fig. 12) attests to the relative unavailability of a response that can compete effectively with the response to the manipulandum.

B. Double Alternation

Traditionally a problem involving two spatial discriminanda, the procedure of double alternation is like that of single alternation except for the sequence of rewarded positions. To obtain reward on every trial the subject must respond in either a RRLL . . . or a LLRR . . . sequence. Experimental sessions are often divided into a number of separate "problems," each consisting of one or more modules of four consecutive trials. The temporal intervals between the problems are longer than those between the trials within a problem.

Double alternation has been believed difficult to acquire because of the inconsistency in relations between antecedent and subsequent events. For example, a response to the right on trial 1 must be followed by a like response on trial 2, while a response to the right on trial 2 must be followed by a different response on trial 3. If an animal is to solve the problem, its behavior must be determined by events more remote than those on the immediately previous trial.

The literature on double alternation lacks balance. Since the introduction of the problem by Hunter (1920), many researchers have tried to demonstrate the degree of proficiency attainable by members of one or another species. As indicated in Chapter 7 by Warren, little evidence supports either the notion of a hierarchy of ability on this problem or the notion that "higher mental processes" are required for solution. Careful parametric studies designed to identify the sources of difficulty in double alternation have been conspicuously absent. However, some analyses of response tendencies in double-alternation settings have begun to provide valuable data.

Just as animals show "hypotheses" in discrimination learning (see Chapter 3 by Levine), they respond systematically in various ways long before they solve double alternation. A subject is typically trained on only one kind of 4-trial problem at a time; under this method at least, the appropriate first response of a run is learned at an early stage by raccoons (Johnson, 1961) and cats (Stewart & Warren, 1957). Rhesus monkeys often adopt a single-alternation hypothesis, which yields a +−−+ pattern of outcomes, or a win-stay, lose-shift hypothesis, which produces a ++−+ pattern (Leary *et al.,* 1952). The latter hypothesis, though less than perfect in its yield of rewards, is considerably better than could be expected by random responding. This suggests that if percentage of correct responses is the criterion of learning, some value well above 75% must be used to assure that double alternation has in fact been acquired.

Yamaguchi and Warren (1961) trained cats on 4-trial problems, and found significant amounts of positive transfer from single to double alternation and from double to single alternation. In both instances the savings were about 40%. The experimenters attributed facilitation to the identity of the initial and terminal location of reward in the two alternation sequences, RLRL and RRLL. Patterns of errors supported their interpretation. In the original learning of either problem, the errors were at first distributed evenly over the four trials. Subjects going from single to double alternation made few errors on trial 1 from the very start. Subjects going from double to single alternation performed well initially on both trials 1 and 4, but quite poorly on trial 3. The difficulty of defining an adequate criterion for double-alternation learning was remarkably well illustrated in this experiment. Original training on both alternation tasks was carried to a criterion of 80% correct in five consecutive daily sessions, each consisting of 10 four-trial problems. Under these conditions, mean days to criterion were 49 for double alternation and 89 for single alternation. After the transfer tasks discussed above, animals of each group were retrained on their initial tasks, and then the number of trials per problem was progressively extended. Single alternation was generalized with little difficulty to problems of 6, 8, 10, 12, 16, and 20 trials. Extension of double alternation to 6-trial problems took about twice as long as original learning, and its further extension to 8-trial problems was not possible for any subject within 60 more days of training. The criteria that had been set for original learning and relearning were sufficient to assure generalizability of single, but not of double, alternation.

A satisfactory criterion for double-alternation learning has not been determined. Greater generalizability might be achieved if the criterion were based on the percentage of perfect intraproblem runs of responses rather than on the percentage of correct responses over all trials. The important unit of performance in double alternation may well be a sequence of responses rather than an individual response. This concept arises from the reasonable (though admittedly unproved) assumption that cue stimuli for a given response may originate as early as the second previous response, if not before. Only as stability is achieved on the earlier response sequences within problems can one expect good performance on later sequences. How can the lengths of single- and double-alternation problems be equated? Again, the answer is not self-evident. Should one keep the number of trials per problem constant? Or should one keep the number of alternations required per problem constant? If the latter course were followed, 4-trial single alternation would be equalled only by 8-trial double alternation. The learning of 8-trial problems from scratch seems at best slow work for rhesus monkeys (Leary *et al.*, 1952) and presumably for other nonhuman subjects. But would correction help? Would some sort of forced-choice procedure help? Would noncontingent information about the locus of reward help? The answers will come only from future research. Even though double alternation may not be as complex as it once was thought to be, it is still complicated enough to present many intriguing problems.

V. SIMILARITY AND DISSIMILARITY PROBLEMS

In the tasks to be discussed here, nonspatial discriminanda are used. Relevant cues derive not from the contiguity of particular discriminanda with particular outcomes, but rather from similarity and dissimilarity relations among the individual discriminanda of stimulus compounds or configurations. Regardless of the specific stimuli present on a trial, a given relation is uniquely associated with one locus of reward. Because consistent choices need not be determined in a narrowly particularistic way, the problems involve conceptual generalization.

A. Sameness-Difference Discrimination

The sameness-difference method introduced recently by J. S. Robinson (1955) features the simultaneous presentation of two compound discriminanda. Each compound consists of a pair of discrete, juxtaposed component sources of stimulation, S. Let us use a hyphen (-) to signify side-by-side pairing. The sources may be identical (e.g., *Sa-Sa, Sb-Sb,* or *Sc-Sc*); or they may be different from one another (e.g., *Sa-Sb, Sa-Sc,* or *Sb-Sc*). One "like" and one "unlike" compound appear on each trial, and "like" pairings are always positive. When an animal has learned to use the relational, sameness-difference cues, it presumably can respond correctly on the very first presentations of new combinations of stimuli. This assumes, of course, that the components of "unlike" pairings are discriminably different to the subject.

In his first study with the task, Robinson (1955) used six chimpanzee subjects. Four had had experience with matching to sample (Section V, D); one had been trained on various object discriminations but not on matching; and one lacked any formal training. A two-foodwell test tray was used. Pairs of objects were mounted on flat rectangular bases. During a sequence of training procedures, the components of the discriminanda were cutout red triangles (*Sa*), blue circles (*Sb*) and black squares (*Sc*). The subjects had to learn to a criterion of 10 consecutive correct responses a trio of problems: *Sa-Sa* versus *Sb-Sc* and its mirror image, *Sc-Sb; Sb-Sb* versus *Sa-Sc* and *Sc-Sa;* and finally *Sc-Sc* versus *Sa-Sb* and *Sb-Sa.* After the animals had mastered these sameness-difference discriminations, they were trained concurrently on all nine possible problems involving different paired combinations of the three components. To assess the effectiveness of his procedures, Robinson followed the concurrent training with a generalization test. Over a series of 40 trials, discriminanda appearing on odd-numbered trials were the same as those that had been used earlier. For even-numbered trials, trial-unique discriminanda were constructed of small manufactured items such as hardware and toys.

Regardless of its learning background, every subject was able to reach criterion on all training procedures. During the generalization test, the percentage of correct choices was 83.3 on the trials with previously used compounds and 77.5 on trials with new compounds. Throughout, the most efficient subjects were those that had had experience at matching to sample. The chimpanzee without behavioral-testing experience solved the first three problems very slowly, and the chimpanzee with only ordinary discrimination-learning experience had especially marked difficulty in reaching the criterion on the concurrent training series.

Interpreting his results, Robinson stressed the generalizability of choices based on sameness-difference relations. Though the experiment was not designed to measure transfer effects, the subjects' very successful performance on test trials indicated facilitation from the training problems. The data are also consistent with the view that matching-to-sample experience aids sameness-difference discrimination. Responsiveness to many irrelevant aspects of the training situation could have impaired the performance of the experimentally naive chimpanzee. Such interference in the absence of prolonged adaptation is not unusual. For the animal with prior discrimination training, several sources of specific interference may have been present. Robinson noted variations in apparatus and in procedural details, and also a difference between the cues relevant to solution in object discrimination and in sameness-difference discrimination. The difference bears some examination.

In the usual object-discrimination problem, the discriminanda look distinctive to human observers and are easily discriminated by nonhuman primates. With "multidimensional" objects differing in several attributes, accurate visual scanning and exact comparison probably are neither demanded nor obtained from subjects. To solve typical stimultaneous object discriminations, for example, rhesus monkeys need not learn relations among stimuli. French (1964) has shown that they can learn to approach a positive object or avoid a negative object in the absence of consistent relational cues. The demonstration involves holding constant over the trials of a problem either the positive or the negative object and pairing with it objects that appear on one and only one trial. It is sufficient that a single object recur and that its association with a particular outcome remain stable. In such an experiment, comparison of the object discriminanda, though possible, is without function. Yet subjects can reach and maintain highly efficient performance. In sameness-difference problems, comparison may be demanded.

Robinson's experiment provides examples of two classes of sameness-difference problem. They differ with respect to the similarity of components in the two compound discriminanda. One may represent the problem-types symbolically in this way:

Type I: *Sa-Sa* versus *Sb-Sc*
Type II: *Sa-Sa* versus *Sa-Sb*

In Type I problems, the identical components of the positive discriminandum are different from either of the components of the negative discriminandum. In the experiment, the first three training problems were of Type I. These problems were probably much like common object-quality discriminations in the observational skills required of the subject. With an animal previously trained on object discrimination, one might look for reasonably good preformance, and the results obtained by Robinson are in line with this expectation. Type II sameness-difference problems are characterized by the negative discriminanda being partial replicas of the positive discriminanda. To make correct responses consistently in Type II problems, the subject must perceive *both* members of at least one compound on each trial. This very likely demands a more thorough "observing response" than is demanded by object-quality discriminations. Problems of Type II were first used by Robinson in the concurrent training series, and it was here that the performance of the chimpanzee previously trained on object discriminations was most dramatically impaired. The subject made over twice as many errors as the next slowest animal, the one that lacked any kind of other training.

Robinson (1960) has proposed that solutions of sameness-difference problems might be based on any of three possible relations: "duplexity," "simplicity," and "multiplicity." The duplexity relation is that the positive discriminandum has two identical components rather than any other number. The simplicity relation is that there are fewer different components in the positive than in the negative discriminandum. The multiplicity relation is that there are more identical elements in the positive than in the negative discriminandum. Among chimpanzees that had been trained extensively on sameness-difference problems, Robinson found by several test procedures that responses were primarily to multiplicity. There was some evidence for responses to simplicity, but none for responses to duplexity.

B. Konorski's Test of Recent Memory

Konorski (1959) has proposed a variant of the sameness-difference problem in which single components are presented successively and do not overlap in time. Depending on whether the two components are alike or differ from one another, dissimilar responses are required. Let us use an ellipsis (. . .) to signify a temporal interval. Compounds *Sa* . . . *Sa* and *Sb* . . . *Sb* might demand a "go" response, while compounds *Sa* . . . *Sb* and *Sb* . . . *Sa* might demand a "no-go" response. Response to the manipulandum must be withheld until the second of the components has been presented. The first component can only determine responses insofar as its organismic effects are retained over some span of time and are available for "comparison" with the organismic effects of the second. The temporal separation of components is assumed to place a burden on the animal's mnemonic capacities.

Stepien and Cordeau (1960) first applied the Konorski method to nonhuman primate subjects. African green monkeys (*Cercopithecus aethiops sabaeus*) were used. During their preliminary adaptation for the experiment, the animals were taught to open the door to a continuously available food-box when a buzzer was sounded and to refrain from opening it when a bell was rung or between trials. This successive discrimination was initially difficult, but training proceeded apace when a mild shock was given for door-openings at inappropriate times. When they had mastered the discrimination, the monkeys were ready for training on an auditory version of the Konorski problem.

The discriminanda for the recent-memory test were two 2-second series of evenly spaced clicks. Within a series, the number of clicks per second was 5 (*Sa*), 10 (*Sb*), or 20 (*Sc*). Stepien and Cordeau originally used two positive compounds, *Sa* . . . *Sa* and *Sc* . . . *Sc*, concurrently with two negative compounds, *Sa* . . . *Sc* and *Sc* . . . *Sa*. A 1-second period intervened between the components. After weeks of training at

50 trials per day, the experimenters tried to aid learning by presenting only one negative compound, *Sc . . . Sa*. Shock was also introduced to inhibit response before the second member of the compound was given. It had become "increasingly clear that the necessity for the animal to hold back its response for some 4–5 sec. after the beginning of each stimulus sequence was one of the greatest obstacles to rapid learning encountered" (Stepien & Cordeau, 1960, p. 390). With the new procedure, the subjects eventually attained 95% correct responses on positive trials and fewer than 5% errors on negative trials. Later, all possible combinations of the three click frequencies were presented, and the interval between stimulus components within a trial was increased to 2, 3, and finally 5 seconds.

During the final phases of training on the auditory task, four of Stepien and Cordeau's seven green monkeys were presented flashes of light at the same time as the clicks and at the same rates. When criterional performance had been attained with the two modes of stimulation combined, the subjects were given only auditory signals during the first 25 trials of a testing session and only visual signals during the last 25 trials. The remaining three monkeys were trained to criterion exclusively with auditory stimuli before the latter procedure was employed. Both groups performed well during the trials on which clicks were used. With the single exception of an animal that appeared to be afraid of the flashing lights, the monkeys had learned to respond correctly with visual signals by the end of the first session. This is a striking contrast to the extended training on the auditory task, which had taken from 33 to 76 testing periods to complete. Since no monkey was trained initially on the visual problem, it is impossible to estimate how much of the great disparity in time of acquisition is attributable to transfer and how much to easier discriminability of the visual cues. One might confidently predict the ability of monkeys as well as other mammals to form learning sets (see Chapter 2 by Miles) for this general type of task. However, no experiment with many sets of discriminanda is available at the time of writing.

Looking back on their use of the Konorski test with green monkeys, Stepien and Cordeau (1960, p. 394) recommend several procedural changes that would preserve the basic attributes of the problem and at the same time possibly decrease or eliminate some of the extraordinary difficulties of training that they encountered. One proposed change is to use a sign-differentiated-position method rather than the successive method. With such a change, *Sa . . . Sa* and *Sb . . . Sb* might be signals to "go left," while *Sa . . . Sb* and *Sb . . . Sa* might be signals to "go right." This two-manipulandum version of the task would allow some response on every trial and make it unnecessary to teach response inhibition. Another suggestion of Stepien and Cordeau is to allow the subject access to the manipulanda only after the second component of the successive discriminandum has been given, rather than throughout the trial. Because the animal's information is not complete until the later component is presented, this change would, without punishment, eliminate premature reactions. A third recommended change in procedure is the use of tones and colors rather than auditory and visual rhythms. This would permit close temporal contiguity among the members of a compound discriminandum. Colors could even be presented simultaneously. Stepien and Cordeau consider that such arrangements might aid earlier stages of training. The plan would be to proceed gradually from simultaneous to successive presentation.

As with the other kinds of problems discussed in this chapter, the test of "recent memory" is broadly modifiable. If the range of applicability of a problem type may be taken as a measure of its analytic utility, Konorski's task seems very valuable. It offers many possibilities for the exploration of interesting temporal parameters. As conceived by Konorski, the problem affords unique opportunities to study separations in time between relevant cue stimuli. In delayed response and delayed alternation, subjects are given all of the cues for a particular trial before the delay. The primary restraint in such problems is on the time of emergence of the outcome-producing response. In the Konorski problem, however, the response need not be postponed when the information is complete. At least a portion of the information may be maximally contiguous with the response.

The essential temporal structure of the Konorski method, as well as its relations to discrimination and delayed response, may be described in terms of four parameters. If S1 and S2 are the first and second members, respectively, of the compound discriminandum, and if R is the response, the parameters are:

$T1$ = duration of S1,
$T2$ = duration of S2,
$(T1 \ldots T2)$ = interval between the beginnings of S1 and S2, and
$(T1 \ldots R)$ = interval between the beginning of S1 and the beginning of the period during which R may be made.

Conditions for the Konorski test are fulfilled whenever $(T1 \ldots T2) > T1$ and $(T1 \ldots R) > (T1 \ldots T2)$. Stepien and Cordeau (1960) suggest that $(T1 \ldots R)$ be at least equal to $(T1 \ldots T2) + T2$. But, as implied before, there is no requirement that the response be delayed until the end of $T2$. Manipulating the parameters, so that $T1 = T2$, $(T1 \ldots T2) = 0$ and $(T1 \ldots R) < T1$, produces either a successive discrimination or a sign-differentiated position discrimination, depending on the nature of the response required. One may then obtain delayed response by increasing ($T1 \ldots R$) so that it is longer than $T1$. It is obviously possible to combine Konorski's task with delayed response by letting $(T1 \ldots T2) > T1$ and $(T1 \ldots R) > (T1 \ldots T2) + T2$. The procedural transformations outlined here would allow some assessment of the relative importance of delays interposed between separate cue elements and those interposed between cue and response.

A great and recurrent question in associative problem-solving is the manner in which organisms can effectively integrate information given to them piecemeal, with gaps of space or time between the fragments. For the study of such separations, the methods proposed by Robinson and by Konorski offer appealingly simple and direct paradigms. Though the amount of research yet accumulated on both types of problem has been small, one may guess that sameness-difference problems will be used increasingly to supplement those of earlier origin. Among the older but still vigorous methods are oddity and matching-to-sample problems, toward which we now turn.

C. Oddity

To obtain reward on each possible occasion in a standard oddity task, an animal need only respond to the odd one among three or more simultaneously-presented discriminanda. Within any trial the positive discriminandum differs in at least one stimulus dimension from the rest, which replicate each other. The various possible combinations and permutations of two kinds of discriminanda in a three-position oddity problem are illustrated in Fig. 1, which shows the six different spatial configurations that can be obtained with horizontal linear arrays. Of these, configurations 1 to 3 contain S_a as the odd object and configurations 4 to 6 contain S_b as the odd object. For any problem in which all six configurations are used, there are trial-to-trial shifts in both the position and the stimulus quality of the odd object. Most commonly the subject is per-

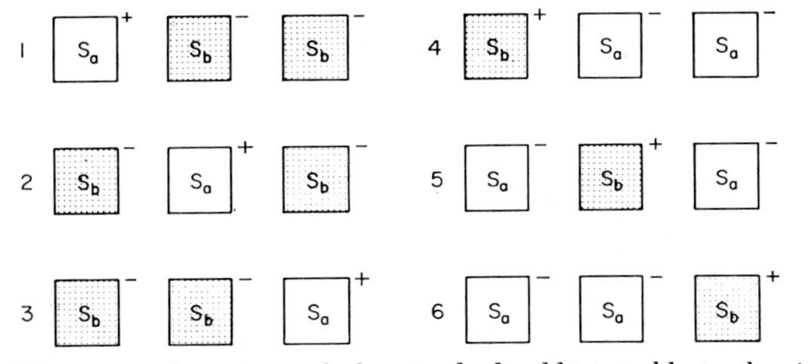

FIG. 1. The six configurations of the standard oddity problem, showing positive (+) and negative (−) discriminanda. S_a and S_b represent dissimilar objects.

mitted one response per trial, and the response may be directed toward any of the three discriminanda.

Levinson (1958) has distinguished between *one-odd problems,* in which only one discriminandum appears as odd, and the more usual *two-odd problems,* in which both discriminanda appear as odd. One-odd problems are discussed in Section II, A, as a form of multiple-choice simultaneous discrimination. Because an animal may learn to respond in terms of oddity relations on the first trials of new problems, even before specific discriminandum-outcome relations can be revealed (e.g., Nissen & McCulloch, 1937b), these problems also truly belong to the oddity group. Consistent responding to oddity implies that a subject can in fact discriminate odd from identical discriminanda, but discriminability does not in itself imply that subjects will use oddity as a basis for response. Some form of oddity training is required.

The one-odd method was first applied in psychophysics. While investigating color discrimination by rhesus monkeys, Kinnaman (1902) carried out some tests with one-odd problems. Klüver (1933, pp. 249–257, 313–320) used the procedure in studying discriminative choice by two cynomolgus monkeys and one cebus monkey, and was impressed by what he believed to be its relative efficiency as contrasted with two-choice discrimination. Both Kinnaman and Klüver only imperfectly appreciated the potentialities of the oddity method as an indicator of an animal's use of relations among discriminanda as a basis of response. Following a plan suggested by Robert C. Tryon, E. W. Robinson (1933) clarified the most basic issue and introduced the two-odd, six-configuration procedure. For the first time, oddity was viewed as a test of "abstraction." In Robinson's study with a cynomolgus monkey, she used the then-popular pulling-in technique (see Chapter 1 by Meyer *et al.*). The animal was required to pull within its reach one of three food boxes, on the fronts of which the discriminanda appeared. The monkey's ability to obtain rewards from the baited box gradually improved in spite of frequent shifts of the odd discriminandum and its location. Throughout her experiment, Robinson used but one set of discriminanda. Possibly the monkey was learning a specific response to each configuration.

It remained for later workers at the University of Wisconsin Primate Laboratory to illustrate the generalizability of oddity responding in multiple problems (Bromer, 1940; Young & Harlow, 1943b). This line of research eventually led Meyer and Harlow (1949) to the concept of oddity-principle-learning set, as shown by the more and more efficient performance of rhesus monkeys on consecutive problems. The subsequent trend toward fewer trials per problem culminated in the demonstration of significant oddity-principle acquisition over many one-trial problems (Levine & Harlow, 1959). This experiment was an especially forceful proof of oddity responding by rhesus monkeys because the animals could not possibly use specific past discriminandum-outcome relations to any consistent advantage. Oddity had reached its full maturity.

One-odd and two-odd problems were directly compared by Levinson (1958, Exp. 1), who changed several features of what had been the usual testing methods. As Moon and Harlow (1955) had done earlier, Levinson used only those four configurations in which the odd stimulus object appears in one of the end positions of a linear array. Configurations 2 and 5 of Fig. 1 were not presented. Levinson required her subjects to touch the nonodd middle object on every trial before the full three-position test tray was made available for response. The "forced choice" was accomplished in the WGTA by means of a centrally-slotted, clear plastic screen. The screen gave the subject manual access to only the middle object until that object had been touched. The transparent screen was then raised to let the subject touch the side objects. Twelve rhesus monkeys, which had had a long history of testing on discrimination-learning problems, were divided into two groups of six for the experiment. Both groups received 12 four-trial oddity problems per day for 22 days. One group was trained throughout on two-odd problems. The other group was given one-odd training for the first 12 days and two-odd training on the remaining 10 days. During the first 12-day period, the percentage of correct responses on trial 1 increased from near chance to a high level in both groups. By this measure, the two-odd group consistently performed better. Within problems, though, the one-odd group improved its standing and the two-odd group deteriorated. As a result, by trial 4 the one-odd animals were making reliably more correct choices than the two-odd animals. The difference between groups on trial 4, large at first, decreased with practice.

How may we understand these findings? It must be recognized immediately that Levinson's forced-choice procedure cannot provide any clear proof of oddity responding. The subject may always successfully just "shift" its response with respect to the object in the central position and may never learn to respond to oddity. That Levinson's rhesus monkeys had in fact learned to use oddity as a cue was shown only in a later study (Levinson, 1958, Exp. 2). At that time, the same animals showed a very substantial amount of transfer to oddity problems not involving the forced-choice technique.

Accepting the reservation that lose-shift responding is always appropriate in her method, Levinson hypothesized a multiple basis for correct responding in her experiment. Within one-odd problems, of course, responding in terms either of oddity or of specific discriminandum-outcome relationships can lead to success. Within two-odd problems oddity is always present as a cue, but discriminandum-outcome relationships fluctuate. The correct object on one trial can become incorrect on the next, a kind of reversal not found in one-odd problems. Especially for subjects trained in discrimination, one might expect some progressive decrease in correct responding within two-odd problems. This did occur early in two-odd training and became less important later. The gradual disappearance of the within-problem decrement indicates that the subjects were learning to use other, more reliable cues.

Still unexplained is the persistent superiority of the two-odd group on the first trials of problems. Two-odd training may offer an advantage in its exposure of subjects to twice as many odd objects as one-odd training. Development of generalized responses to oddity may occur quicker in animals that are presented with a wider range of odd discriminanda. This possibility remains to be tested experimentally.

In Levinson's first experiment, introducing two-odd training after 12 days of one-odd training did not impair performance on trial 1 but reduced efficiency drastically at first on the later trials of problems. This finding supports the position that intraproblem changes of odd objects can be expected to interfere with the performance of animals that have acquired a win-stay, lose-shift hypothesis. An identical conclusion arises with much firmer conviction from the data of Levine and Harlow (1959). Again, rhesus monkeys had been trained on discrimination before the investigation began. Different groups were given 1-trial and 12-trial oddity problems. From the very beginning, total performance was markedly better in the 1-trial-per-problem group. Levine and Harlow's detailed analysis of the response sequences of animals of the 12-trial-per-problem group leaves no doubt of the importance of "object-discrimination responding" in determining their choices (see Chapter 3 by Levine). When this group was given a series of 1-trial problems at the end of the experiment, its over-all efficiency immediately approximated that of the other group.

In the four-configuration oddity problems used by Levinson (1958, Exp. 1), the central discriminandum of a display conveys essential information in spite of its always being negative. Levinson's forced-choice procedure assures that on each trial the subject's hand, and more importantly its eye, is directed toward the center object. Although this technique does not guarantee that the animal will perceive the middle object, it certainly facilitates the appropriate observation (see Chapter 1 by Meyer *et al.*). The arrangements used by Levinson are reminiscent of those used very successfully by Skinner (1950) and by Ginsburg (1957) to teach pigeons matching (and nonmatching) to sample. Levinson's procedure was more effective than those of other recent experimenters on oddity. Moon and Harlow (1955) allowed their rhesus monkeys to move any one of the three objects on the trials of two-odd, four-configuration problems. Displacement of the never-positive central object was allowed to extinguish during the course of training. This method, though less effective than Levinson's, gave much faster learning than that of Levine and Harlow (1959), who presented the nonodd center object always behind a transparent plastic barrier, which kept animals from ever touching that object. Even under Levine and Harlow's easier, one-trial-per-problem condition, the percentage of correct responses was only slightly greater than 70 after some 3,000 trials. In contrast, Moon and Harlow's subjects reached 88% correct in a total of 1,150 trials. Levinson's two-odd group reached 95% correct in fewer than 500 trials. It seems imperative that monkeys in oddity experiments eventually develop some display-scanning acts. Such observing responses are the basis for successful performance in procedures which demand comparison of simultaneously-presented discriminanda. Any procedure that promotes manipulation in regions of the apparatus from which relevant cues may be obtained would appear to be helpful in acquisition.

Lockhart and Harlow (1962) directly compared oddity-learning-set formation in groups of pig-tailed monkeys (*Macaca nemestrina*) trained on either standard six-configuration oddity or four-configuration oddity with response to the center position never rewarded. During 600 four-trial problems, the animals were allowed unrestricted access to all positions of objects. A noncorrection procedure was followed. The over-all difference in the percentage of correct responses under the two conditions was not reliable. After their monkeys had attained high accuracy, Lockhart and Harlow measured latencies. Again, there was no reliable difference between groups. At least on global measures of performance, whether or not the odd object ever occupied the middle position was of little moment.

The relation between performance and the distribution of objects within stimulus displays is a topic of nascent interest. For six-configuration problems, rhesus monkeys frequently make more errors when the odd object is in the center position than when it is at either end (e.g., Levinson, 1958; Odoi, 1956). That the effect is sometimes absent (Riopelle, 1959) emphasizes the current lack of detailed information concerning it. Lockhart and Harlow (1962) found that center-position difficulty increases as acquisition proceeds. The same seems also to be true for end positions relatively close to the center as opposed to those that are relatively distant from the center. In earlier stages of training, efficiency is enhanced by more-compact groupings of objects. Are these secondary effects of a change in display-scanning habits during oddity training?

Nonspatial discriminanda in associative problems have nearly always been presented along a horizontal straight line in the frontal plane of the subject. Riopelle (1959) examined the effects on rhesus monkeys of nonlinear arrangements of oddity discriminanda in the same plane. His subjects had been trained previously in the usual manner. In Riopelle's tests, objects were dispersed both horizontally and vertically in several types of triangular displays. Any departure from the linear arrangement reliably reduced the proportion of correct responses. Although the decrement may have been due to the animals' earlier development of circumscribed habits of observation, possibly even naive monkeys would have more difficulty with nonlinear than with linear arrays.

Bernstein (1961) has introduced a promising variation of oddity in the *dimension-abstracted* problem. The discriminanda are sets of objects. Within a single relevant stimulus dimension, all negative objects differ from the positive object in the same way and to the same extent. But no two objects appearing on a given trial need be exactly alike. The negative objects may differ both from the positive object and from one another in many irrelevant stimulus dimensions. "Only negative objects can be matched to each other, and only by using the relevant cue" (Bernstein, 1961, p. 247).

In his experiments, Bernstein used five-position problems of the kinds exemplified in Fig. 2. His subjects included human adults, chimpanzees, an orangutan (*Pongo pygmaeus*), pig-tailed monkeys, and rhesus mon-

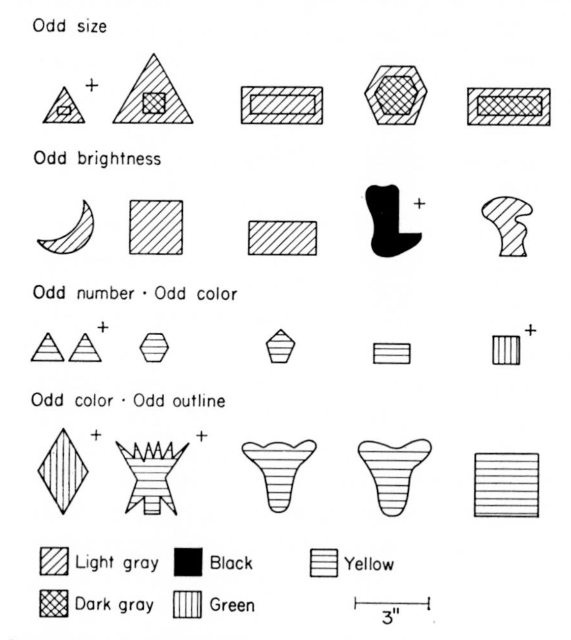

FIG. 2. Four dimension-abstracted oddity problems. Plus signs indicate positive discriminanda. (After Bernstein, 1961.)

keys. In the first part of the research, each problem had one relevant stimulus dimension, either size or brightness. Trained concurrently on a list of 70 problems, the apes reached a 90% criterion earlier than the monkeys. Between the two species of ape and the two species of monkey, there were no significant differences in performance. In later problems with two relevant dimensions, monkey and human subjects could obtain reward by moving either of two objects. This allowed tests of differential strength of various potential cues. The relevant dimensions were color and 36 kinds of variation in form—e.g., external shape, symmetry, number, curvature, interruption of outline, presence of a hole, and relative lengths of axes. In general, humans and macaques showed similar preferences for certain dimensions. Color was dominant in many, but not all, problems. Much depended on the precise nature of the several stimuli presented together. Bernstein concluded that no sweeping generalization about the relative importance of color and form cues was justified. But Bernstein's method seems to be a practical new way of assessing the nature of the effective stimuli in associative problems.

D. Matching to Sample

The matching-to-sample problem is a logical companion to oddity. They are in many ways the inverse of one another. In matching, the animal must react positively to sameness rather than difference. A trial is split into two phases: sampling and choice. During sampling, the animal responds in some way defined by the experimenter to a single discriminandum. If the sample is an object, it may be segregated in some part of the test tray, and the response may consist of pushing it aside, followed by immediate receipt of food. With the sample still present for comparison, a choice then becomes possible. Two or more additional objects, at least one of which is like the sample, are made accessible. To obtain more food, the subject must displace the replica (or replicas) of the sample.

As an example of the use of matching to sample, let us consider the successful attempt by Weinstein (1945) to teach color-categorizing to rhesus monkeys. Historical precedent for the use of this method to produce generalized concept learning had been established by Itard's work during the early 19th Century with the "wild boy" of Aveyron (Itard, 1932). Weinstein's research was a truly remarkable demonstration of the ability of nonhuman primates to form abstract concepts. His experiment may be compared with the studies of Andrew and Harlow (1948) on shape concepts, of Hicks (1956) on number concepts, and of Klüver (1933, pp. 129–172) on size concepts. Whereas the other investigators approached generalization through increasing breadth of simultaneous discrimination training, Weinstein used a stepwise matching technique to establish greater and greater independence of choice from irrelevant stimulus dimensions and from specific procedural details.

Weinstein's subjects were two rhesus monkeys, Corry and Zo. Corry had had extensive laboratory experience and was considerably less excitable during testing. In the course of their stepwise training, the subjects had to reach criterional performance at any given stage before being advanced to the next. The animals began by matching identically-colored objects. During this preliminary stage, they progressed from dealing with one set of discriminanda to dealing with multiple sets. They also progressed from selecting one of two choice objects to selecting one of four.

Color categorizing began with two cutout wooden samples: red triangle and blue ellipse. Corry and Zo had to select among the choice objects on the basis of color identity, regardless of form or size. Next, they were trained to select 0, 1, 2, or 3 positive choice objects *and* to refrain for 30 seconds from choosing any negative object. Although this stage of acquisition presented some difficulties, particularly in the required inhibition of responses to negative objects, both subjects mastered it. They were then advanced to sorting for hue in spite of variations of brightness and saturation, and to sorting among combinations of thick and thin objects. An especially trying stage followed. In it, the choice objects were plaques of varied surface patterns, the differential color cues being confined to areas in the centers. Zo became very disturbed at this stage and had to be dropped from the experiment, but Corry met the criterion and went on to attain some of the most outstanding conceptual performances ever reported for monkeys.

Up to this point Corry had been required to choose from among four objects arrayed in a straight line. A deeper test tray was introduced, and the animal now had to sort among various configurations of as many as eight objects. Sometimes he had to reach over negative objects to displace positive objects (see Fig. 10 in Chapter 1 by Meyer *et al.*). During the same stage of training, the sample was displayed in the experimenter's hand at various locations above the tray or was given to the animal to hold. None of these procedural changes permanently impaired Corry's sorting ability. Then the cues provided by the two sample objects were reduced and color was eliminated. Corry now learned to choose red objects when an unpainted triangle was presented and to choose blue objects when an unpainted ellipse was presented. Finally, sample and matching objects were successfully interchanged. The monkey learned to select triangles whenever red objects were presented and to select ellipses whenever blue objects were presented.

Several of Weinstein's results are quite impressive. Harlow (1951) has emphasized that "*not responding* to stimuli that lay outside the category was as important an aspect of the problem as was responding to stimuli within the category" (Harlow, 1951, p. 234, italics his). Nonresponse to negative objects showed that the color categories had become defined exclusively as well as inclusively. Nearly as striking is the use of "samples" in which the critical cue was not directly represented. One should not assume that the capability for doing this is the unique possession of primates, or even mammals. Ginsburg (1957) has obtained similar "amatching" in homing pigeons. The analogy between amatching and the human use of words to designate classes of nonverbal attributes is obvious.

Although the matching procedure ordinarily provides excellent opportunities for comparison of sample and choice discriminanda, some minimal delay between the perceptions of spatially separate discriminanda is inevitable. If the experimenter removes the sample after the sampling phase of a trial and extends the interval before choice is possible, we have a *delayed matching* problem. Following training on regular matching, the introduction of delays at least temporarily interferes with performance of rhesus monkeys using object discriminanda (Weinstein, 1941) and of chimpanzees using color discriminanda (Riesen & Nissen, 1942). Mishkin *et al.* (1962) have shown that, with 5-second delays, rhesus monkeys eventually can perform very successfully on matching or nonmatching to either baited or unbaited samples. Finch (1942) has compared delayed matching with a form of delayed response in which the to-be-positive object is simply displayed, but not with food, before the delay. Chimpanzees perform much better on delayed matching than on Finch's nonspatial delayed response.

Spaet and Harlow (1943a) employed a progressive training method to obtain delayed matching concurrently from baited and unbaited samples by rhesus monkeys. Trained first on the usual simultaneous matching, the subjects were next given *interrupted matching*, in which the opaque screen of the WGTA was lowered between sampling and choice phases but with the sample still present during the choice phase. Only very mild and transitory disruption occurred. Spaet and Harlow then introduced what they considered to be a matching analog of spatial delayed response. In it, both sample and matching objects were present during sampling, allowing subjects to identify the spatial locus of the positive choice object at that time. Later the monkeys were trained on a matching analog of nonspatial delayed response, in which the choice objects appeared only after the delay. Finally, reward was given on only half the trials at random under the nonspatial condition. The subjects, performing above chance level under all conditions, were apparently able to maintain a "stay" strategy with respect to the sample in the absence of 100% reward.

As long as matching is conducted with a limited number of sample and choice objects which recur trial after trial, successful matches may be attributed to the acquisition of specific associations. Generalized matching with new sets of discriminanda has been demonstrated effectively with chimpanzees (Nissen *et al.*, 1948) and with rhesus monkeys (Mishkin *et al.*, 1962). Since subjects with appropriate training can match successfully after one exposure to a sample, we may speak of the development of matching-to-sample learning sets.

There are obvious similarities between matching with unique discriminanda and two-trial simultaneous-discrimination problems. The parallels are not complete, however. In discrimination, the animal must make a choice on the first trial, expressing its preference for one of the discriminanda. In matching, the animal responds to the sample regardless of preference. Also, while a "win-stay, lose-shift" strategy is almost always appropriate in discrimination problems, a simple strategy of "stay" may be completely adequate in matching. As noted above, good matching performance can be obtained with either "win" or "lose" outcomes of responses to the sample.

E. Higher-Order Conditional Problems

There is an element of conditionality in all similarity and dissimilarity problems. It lies in the trial-specificity of the relationships between discriminanda and outcomes. Beyond that, one may impose other kinds of conditionality, creating additional variants of the standard problems.

One variant is the *Weigl-principle oddity* problem of Young and Harlow (1943a, 1943b). It was derived from the sorting test invented by Weigl (1941) as an indicator of the capacity of humans to classify blocks by color or shape and to shift rapidly among several sets of categories. In this form of oddity, one object may be odd in shape and another odd in brightness (see Fig. 3). Depending on the color of the test tray, either odd shape or odd brightness is associated with reward. Young and Harlow (1943a) showed that solving problems similar to this is well within the ability of rhesus monkeys. The solutions can be generalized to problems involving novel sets of stimuli, to problems with more than three objects, and to problems with similar but nonidentical negative discriminanda (Young & Harlow, 1943b).

Weigl-principle matching problems were introduced by Harlow (1943). He successfully required rhesus monkeys to match for color following reward of response to a sample and to match for form following nonreward of response to a sample. Nissen *et al.* (1948) trained chimpanzees on a problem in which the conditional cue was given by test-tray color and in which no displacement of a centrally-located "sample" was demanded. On this modification of Weigl matching, the chimpanzees were greatly inferior to Harlow's monkeys in the accuracy of their performance. Harlow (1951) interpreted the difference as depending upon the diversity of "personality characteristics" between the two species. Students today might be more inclined to indict the apparatus and procedure of Nissen and his co-workers. It is at least as likely that the chimpanzees' poor showing was due to the perceptual difficulties posed by the method as it was to their sullen and refractory natures.

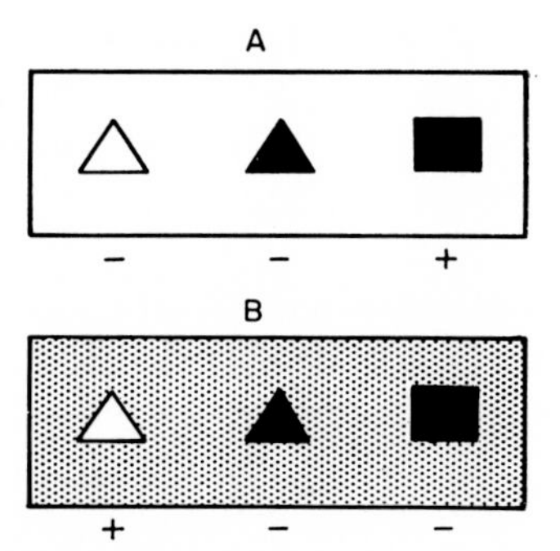

Fig. 3. Weigl-principle oddity problem. On test tray A, odd shape is positive. On test tray B, odd brightness is positive. The conditional cue is provided by the background on which discriminanda appear. Other configurations of the objects (not shown here) are possible in the same problem.

Another modification of standard oddity is the combined *oddity-nonoddity* problem, in which a conditional discriminandum indicates the object or objects associated with reward on a given trial. In the experiments by Spaet and Harlow (1943b) and by Harlow and Moon (1956), the conditional cue was the color of the test tray. Figure 4 illus-

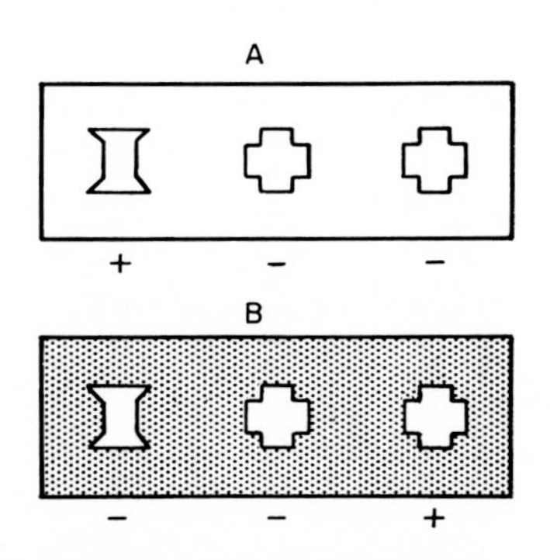

Fig. 4. Oddity-nonoddity problem. On test tray A, the odd object is positive. On test tray B, the nonodd object in the end position is positive. In this version of the problem, the center object is always nonodd and negative. If the center object is regarded as a sample object to which nonresponse is required, this problem is also a matching-nonmatching problem: matching on test tray B, nonmatching on test tray A.

trates Harlow and Moon's procedure, in which the nonodd center object never covers food. Whenever one test tray is used, the odd object covers food. Whenever the other tray is used, the nonodd object in an end position covers food. Spaet and Harlow's procedure differed in two ways. Food might be found from trial to trial at any one of the three positions, and in the presence of the cue for nonoddity both nonodd objects covered food. A response to nonoddity was counted when both like objects were chosen with no preceding or intervening response to the odd object. Under either set of conditions and with the perseverance of subjects and experimenters, rhesus monkeys can be taught oddity-nonoddity.

Matching-nonmatching problems may be coupled in the same fashion as oddity-nonoddity problems. Harlow (1942) and Harlow and Spaet (1943) showed that rhesus monkeys can use the presence or absence of food under a sample object as a conditional cue for matching or nonmatching. Under the conditions of training in Harlow's experiment, a high criterion is achievable in 1,000 to 2,000 trials. Depending on the procedure elected by an experimenter, either matching or nonmatching can be cued by a baited or an unbaited sample.

A moment's reflection will serve to convince the reader of the near-identity of matching and nonoddity and of nonmatching and oddity (see Fig. 4). The most striking difference is the response to the sample required in the matching procedures. But the response may be present in some modifications of oddity (Levinson, 1958), and may be absent in some modifications of matching (Nissen *et al.*, 1948).

Harlow (1942) provided the literature on associative problems with its outstanding illustration of the capacity of rhesus monkeys to achieve mastery of three concurrently-presented conditional problems. These were a matching-nonmatching problem, a sign-differentiated position discrimination (Section II, C), and a reversed sign-differentiated position discrimination. The same two kinds of objects (push buttons and crosses) appeared throughout the experiment, each of the trials of which was run under a two-part, sample-and-choice procedure. When the choice objects were different from one another, matching or nonmatching was required. When the choice objects were identical to one another but different from the sample, the original sign-differentiated position discrimination was required. When the choice objects were identical both to one another and to the sample, the reversed sign-differentiated position discrimination was required. Within each major category of problem, the appropriate choice was signaled by reward or nonreward of response to the sample. Using a stepwise procedure, Harlow first trained the animals on individual conditional problems and then on regular and finally on random alternations among problems. The four subjects (which had previously learned matching-nonmatching cued by tray color) attained the seven successive criteria of training within 1,100 to 2,400 total trials, impressively small figures in comparison to the many thousands of trials that are sometimes expended to achieve a lesser result.

VI. CONCLUSIONS

This chapter has emphasized the methodology of associative problem-solving tasks and the relationships among members of this family of problems. Similarities and differences in the structural aspects of discriminandum-outcome associations have been described for various problems. The effectiveness of choice behavior in producing reward may be related to several variables: the characteristics of the stimulus display, trial-to-trial regularities in the location of rewards, opportunities for comparison among discriminanda, requirements of directional variation of response or inhibition of response, and the subject's experience with problems of the same or similar types. Although the associative problems discussed here may indeed contain complexities, the methods of identifying the complexities and of assessing their true importance leave much to be desired. Complexities are frequently ephemeral. Once highly tamed and stable animals have formed the necessary learning sets, they can solve most of the major classes of problem discussed here in the absolute minimum number of trials needed to convey definitive information about a given discriminandum-outcome association. In discrimination learning, the minimum is one trial. In similarity and dissimilarity problems, no prior trials with a new set of discriminanda are needed by a fully-trained animal.

If one may extrapolate from the past, the future of research on associative problems will profitably concentrate on the determinants of problem solution, whether they lie in the history of the subject or in the structure of the immediate task. There is every indication that the vigor of experimental work in this area will be sustained for many years.

REFERENCES

Andrew, Gwen, & Harlow, H. F. (1948). Performance of macaque monkeys on a test of the concept of generalized triangularity. *Comp. Psychol. Monogr.* **19**, No. 3 (Serial No. 100).

Behar, I. (1961). Analysis of object-alternation learning in rhesus monkeys. *J. comp. physiol. Psychol.* **54**, 539.

Bernstein, I. S. (1961). The utilization of visual cues in dimension-abstracted oddity by primates. *J. comp. physiol. Psychol.* **54**, 243.

Bromer, J. A. (1940). A genetic and physiological investigation of concept-behavior in primates. Unpublished doctoral dissertation, University of Wisconsin.

Brown, W. L., & McDowell, A. A. (1963). Response shift learning set in rhesus monkeys. *J. comp. physiol. Psychol.* **56**, 335.

Bush, R. R., & Mosteller, F. (1955). "Stochastic Models for Learning." Wiley, New York.

Cotterman, T. E., Meyer, D. R., & Wickens, D. D. (1956). Discrimination reversal learning in marmosets. *J. comp. physiol. Psychol.* **49,** 539.

Crawford, F. T. (1962). Reversal learning to spatial cues by monkeys. *J. comp. physiol. Psychol.* **55,** 869.

Dufort, R. H., Guttman, N., & Kimble, G. A. (1954). One-trial discrimination reversal in the white rat. *J. comp. physiol. Psychol.* **47,** 248.

Finch, G. (1942). Delayed matching-from-sample and non-spatial delayed response in chimpanzees. *J. comp. Psychol.* **34,** 315.

French, G. M. (1959). Performance of squirrel monkeys on variants of delayed response. *J. comp. physiol. Psychol.* **52,** 741.

French, G. M. (1964). The frontal lobes and association. *In* "The Frontal Granular Cortex and Behavior" (J. M. Warren & K. Akert, eds.), pp. 56–73. McGraw-Hill, New York.

Ginsburg, N. (1957). Matching in pigeons. *J. comp. physiol. Psychol.* **50,** 261.

Gollin, E. S., & Liss, P. (1962). Conditional discrimination in children. *J. comp. physiol. Psychol.* **55,** 850.

Gross, C. G. (1963a). A comparison of the effects of partial and total lateral frontal lesions on test performance by monkeys. *J. comp. physiol. Psychol.* 56, 41.

Gross, C. G. (1963b). Discrimination reversal after lateral frontal lesions in monkeys. *J. comp. physiol. Psychol.* **56,** 52.

Gross, C. G. (1963c). Effect of deprivation on delayed response and delayed alternation performance by normal and brain operated monkeys. *J. comp. physiol. Psychol.* **56,** 48.

Gross, C. G., & Weiskrantz, L. (1962). Evidence for dissociation of impairment on auditory discrimination and delayed response following lateral frontal lesions in monkeys. *Exp. Neurol.* **4,** 453.

Harlow, H. F. (1942). Responses by rhesus monkeys to stimuli having multiple sign-values. *In* "Studies in Personality" (Q. McNemar & Maud A. Merrill, eds.), pp. 105–123. McGraw-Hill, New York.

Harlow, H. F. (1943). Solution by rhesus monkeys of a problem involving the Weigl principle using the matching-from-sample method. *J. comp. Psychol.* **36,** 217.

Harlow, H. F. (1950a). Analysis of discrimination learning by monkeys. *J. exp. Psychol.* **40,** 26.

Harlow, H. F. (1950b). Performance of catarrhine monkeys on a series of discrimination reversal problems. *J. comp. physiol. Psychol.* **43,** 251.

Harlow, H. F. (1951). Primate learning. *In* "Comparative Psychology" (C. P. Stone, ed.), 3rd ed., pp. 183–238. Prentice-Hall, New York.

Harlow, H. F., & Moon, L. E. (1956). The effects of repeated doses of total-body X radiation on motivation and learning in rhesus monkeys. *J. comp. physiol. Psychol.* **49,** 60.

Harlow, H. F., & Spaet, T. (1943). Problem solution by monkeys following bilateral removal of the prefrontal areas. IV. Responses to stimuli having multiple sign values. *J. exp. Psychol.* **33,** 500.

Hayes, K. J., Thompson, R., & Hayes, Catherine (1953). Concurrent discrimination learning in chimpanzees. *J. comp. physiol. Psychol.* **46,** 105.

Hicks, L. H. (1956). An analysis of number-concept formation in the rhesus monkey. *J. comp. physiol. Psychol.* **49,** 212.

House, Betty J., & Zeaman, D. (1959). Position discrimination and reversals in low-grade retardates. *J. comp. physiol. Psychol.* **52,** 564.

Hunter, W. S. (1920). The temporal maze and kinesthetic sensory processes in the white rat. *Psychobiology* **2,** 1.

Itard, J.-M.-G. (1932). "The Wild Boy of Aveyron" (Transl. G. Humphrey & Muriel Humphrey.) Century, New York.

Johnson, J. I., Jr. (1961). Double alternation in raccoons. *J. comp. physiol. Psychol.* **54,** 248.

Kinnaman, A. J. (1902). Mental life of two *Macacus rhesus* monkeys in captivity. —I. *Amer. J. Psychol.* **13,** 98.

Klüver, H. (1933). "Behavior Mechanisms in Monkeys." Univer. Chicago Press, Chicago, Illinois.

Konorski, J. (1959). A new method of physiological investigation of recent memory in animals. *Bull. Acad. Polonaise Sci.* **7,** 115.

Ladieu, G. (1944). The effect of length of the delay interval upon delayed alternation in the albino rat. *J. comp. Psychol.* **37,** 273.

Leary, R. W. (1957). The effect of shuffled pairs on the learning of serial discrimination problems by monkeys. *J. comp. physiol. Psychol.* **50,** 581.

Leary, R. W., Harlow, H. F., Settlage, P. H., & Greenwood, D. D. (1952). Performance on double-alternation problems by normal and brain-damaged monkeys. *J. comp. physiol. Psychol.* **45,** 576.

Levine, M., & Harlow, H. F. (1959). Learning-sets with one- and twelve-trial oddity-problems. *Amer. J. Psychol.* **72,** 253.

Levinson, Billey (1958). Oddity learning set and its relation to discrimination learning set. Doctoral dissertation, University of Wisconsin. University Microfilms, Ann Arbor, Michigan, No. 58–7507.

Lockhart, J. M. & Harlow, H. F. (1962). The influence of spatial configuration and percentage of reinforcement upon oddity learning. *J. comp. physiol. Psychol.* **55,** 495.

Loucks, R. B. (1931). Efficacy of the rat's motor cortex in delayed alternation. *J. comp. Neurol.* **53,** 511.

McClearn, G. E. (1957). Differentiation learning by monkeys. *J. comp. physiol. Psychol.* **50,** 436.

McCulloch, T. L., & Nissen, H. W. (1937). Equated and non-equated stimulus situations in discrimination learning by chimpanzees. II. Comparison with limited response. *J. comp. Psychol.* **23,** 365.

McDowell, A. A., & Brown, W. L. (1963a). The learning mechanism in response shift learning set. *J. comp. physiol. Psychol.* **56,** 572.

McDowell, A. A., & Brown, W. L. (1963b). Learning mechanism in response perseveration learning sets. *J. comp. physiol. Psychol.* **56,** 1032.

Meyer, D. R. (1951). Food deprivation and discrimination reversal learning by monkeys. *J. exp. Psychol.* **41,** 10.

Meyer, D. R., & Harlow, H. F. (1949). The development and transfer of response to patterning by monkeys. *J. comp. physiol. Psychol.* **42,** 454.

Mishkin, M., & Pribram, K. H. (1955). Analysis of the effects of frontal lesions in monkey: I. Variations of delayed alternation. *J. comp. physiol. Psychol.* **48,** 492.

Mishkin, M., & Pribram, K. H. (1956). Analysis of the effects of frontal lesions in monkey: II. Variations of delayed response. *J. comp. physiol. Psychol.* **49,** 36.

Mishkin, M., Prockop, Elinor S., & Rosvold, H. E. (1962). One-trial object-discrimination learning in monkeys with frontal lesions. *J. comp. physiol. Psychol.* **55,** 178.

Moon, L. E., & Harlow, H. F. (1955). Analysis of oddity learning by rhesus monkeys. *J. comp. physiol. Psychol.* **48,** 188.

Morgan, C. T. (1961). "Introduction to Psychology," 2nd ed. McGraw-Hill, New York.

Nissen, H. W. (1942). Ambivalent cues in discriminative behavior of chimpanzees. *J. Psychol.* **14,** 3.

Nissen, H. W. (1951). Analysis of a complex conditional reaction in chimpanzee. *J. comp. physiol. Psychol.* **44,** 9.

Nissen, H. W., & McCulloch, T. L. (1937a). Equated and non-equated stimulus situations in discrimination learning by chimpanzees. I. Comparison with unlimited response. *J. comp. Psychol.* **23,** 165.

Nissen, H. W., & McCulloch, T. L. (1937b). Equated and non-equated stimulus situations in discrimination learning by chimpanzees. III. Prepotency of response to oddity through training. *J. comp. Psychol.* **23,** 377.

Nissen, H. W., Blum, Josephine S., & Blum, R. A. (1948). Analysis of matching behavior in chimpanzee. *J. comp. physiol. Psychol.* **41,** 62.

Noer, Mary C., & Harlow, H. F. (1946). Discrimination of ambivalent cue stimuli by macaque monkeys. *J. gen. Psychol.* **34,** 165.

North, A. J., Maller, O., & Hughes, C. (1958). Conditional discrimination and stimulus patterning. *J. comp. physiol. Psychol.* **51,** 711.

Odoi, H. (1956). Analysis of the factors involved in oddity performance of monkeys. *Proc. S. Dak. Acad. Sci.* **35,** 220.

Osgood, C. E. (1953). "Method and Theory in Experimental Psychology." Oxford Univer. Press, New York.

Pribram, K. (1959). On the neurology of thinking. *Behav. Sci.* **4,** 265.

Pribram, K. H., & Mishkin, M. (1955). Simultaneous and successive visual discrimination by monkeys with inferotemporal lesions. *J. comp. physiol. Psychol.* **48,** 198.

Pribram, K. H., & Mishkin, M. (1956). Analysis of the effects of frontal lesions in monkey: III. Object alternation. *J. comp. physiol. Psychol.* **49,** 41.

Riesen, A. H., & Nissen, H. W. (1942). Non-spatial delayed response by the matching technique. *J. comp. Psychol.* **34,** 307.

Riley, D. A., Ring, K., & Thomas, J. (1960). The effect of stimulus comparison on discrimination learning and transposition. *J. comp. physiol. Psychol.* **53,** 415.

Riopelle, A. J. (1959). Linear and nonlinear oddity. *J. comp. physiol. Psychol.* **52,** 571.

Riopelle, A. J., & Copelan, E. L. (1954). Discrimination reversal to a sign. *J. exp. Psychol.* **48,** 143.

Riopelle, A. J., & Francisco, E. W. (1955). Discrimination learning under different first-trial procedures. *J. comp. physiol. Psychol.* **48,** 90.

Robinson, E. W. (1933). A preliminary experiment on abstraction in a monkey. *J. comp. Psychol.* **16,** 231.

Robinson, J. S. (1955). The sameness-difference discrimination problem in chimpanzee. *J. comp. physiol. Psychol.* **48,** 195.

Robinson, J. S. (1960). The conceptual basis of the chimpanzee's performance on the sameness-difference discrimination problem. *J. comp. physiol. Psychol.* **53,** 368.

Rumbaugh, D. M., & McQueeney, J. A. (1963). Learning-set formation and discrimination reversal: Learning problems to criterion in the squirrel monkey. *J. comp. physiol. Psychol.* **56,** 435.

Schusterman, R. J. (1962). Transfer effects of successive discrimination-reversal training in chimpanzees. *Science* **137,** 422.

Schusterman, R. J., & Bernstein, I. S. (1962). Response tendencies of gibbons in single and double alternation tasks. *Psychol. Rep.* **11,** 521.

Skinner, B. F. (1950). Are theories of learning necessary? *Psychol. Rev.* **57,** 193. (Reprinted in "Cumulative Record," pp. 39–69. Appleton-Century-Crofts, New York, 1959).

Spaet, T., & Harlow, H. F. (1943a). Problem solution by monkeys following bilateral removal of the prefrontal areas. II. Delayed reaction problems involving use of the matching-from-sample method. *J. exp. Psychol.* **32,** 424.

Spaet, T., & Harlow, H. F. (1943b). Solution by rhesus monkeys of multiple sign problems utilizing the oddity technique. *J. comp. Psychol.* **35,** 119.

Spence, K. W. (1936). The nature of discrimination learning in animals. *Psychol. Rev.* **43,** 427. (Reprinted in "Behavior Theory and Learning," pp. 269–291. Prentice-Hall, Englewood Cliffs, New Jersey, 1960.)

Spence, K. W. (1939). The solution of multiple choice problems by chimpanzees. *Comp. Psychol. Monogr.* **15,** No. 3 (Serial No. 75).

Spence, K. W. (1952). The nature of the response in discrimination learning. *Psychol. Rev.* **59,** 89. (Reprinted in "Behavior Theory and Learning," pp. 359–365. Prentice-Hall, Englewood Cliffs, New Jersey, 1960.)

Stepien, L. S., & Cordeau, J. P. (1960). Memory in monkeys for compound stimuli. *Amer. J. Psychol.* **73,** 388.

Stewart, C. N., & Warren, J. M. (1957). The behavior of cats on the double alternation problem. *J. comp. physiol. Psychol.* **50,** 26.

Warren, J. M. (1964). Additivity of cues in conditional discrimination learning by rhesus monkeys. *J. comp. physiol. Psychol.* **58,** 124.

Weigl, E. (1941). On the psychology of so-called processes of abstraction. *J. abnorm. soc. Psychol.* **36,** 3.

Weinstein, B. (1941). Matching-from-sample by rhesus monkeys and by children. *J. comp. Psychol.* **31,** 195.

Weinstein, B. (1945). The evolution of intelligent behavior in rhesus monkeys. *Genet. Psychol. Monogr.* **31,** 3.

Weiskrantz, L. (1957). On some psychophysical techniques employing a single manipulandum. *Brit. J. Psychol.* **48,** 189.

Wilson, W. A., Jr. (1962). Alternation in normal and frontal monkeys as a function of response and outcome of previous trial. *J. comp. physiol. Psychol.* **55,** 701.

Wilson, W. A., Jr., & Rollin, A. R. (1959). Two-choice behavior of rhesus monkeys in a noncontingent situation. *J. exp. Psychol.* **58,** 174.

Yamaguchi, S., & Warren, J. M. (1961). Single versus double alternation learning by cats. *J. comp. physiol. Psychol.* **54,** 533.

Young, M. L., & Harlow, H. F. (1943a). Solution by rhesus monkeys of a problem involving the Weigl-principle using the oddity method. *J. comp. Psychol.* **35,** 205.

Young, M. L., & Harlow, H. F. (1943b). Generalization by rhesus monkeys of a problem involving the Weigl-principle using the oddity method. *J. comp. Psychol.* **36,** 210.

Chapter 6
Operant Conditioning[1]

Roger T. Kelleher

Department of Pharmacology, Harvard Medical School, Boston, Massachusetts

I. INTRODUCTION

Although most current behavioral experiments with nonhuman primates use operant conditioning procedures as defined by Skinner (1938), only experiments with certain additional characteristics are consistently referred to as operant conditioning experiments in current psychological terminology (Ferster, 1953). The operant conditioning experiments discussed in this chapter will have these additional characteristics. The first characteristic is the extensive use of rate and pattern of responding as dependent variables. In operant conditioning experiments some response that the organism can repeat readily is usually selected for study. Such responses provide the investigator with a potentially wide range of response rates and response patterns that can be studied as a function of changes in independent variables. The second characteristic is the explicit use of schedules of reinforcement. A schedule of reinforcement is a precise specification of the plan according to which discriminative and reinforcing stimuli will be presented. The use of different schedules

[1]Preparation of this chapter was supported in part by research grants (M-2094, MY-2645, and MH-07658–01) and by a research career development award (K3-MH-22,589–01) from the National Institute of Mental Health, U. S. Public Health Service. I am indebted to Dr. P. B. Dews and Dr. W. H. Morse for their helpful comments during the preparation of this chapter.

of reinforcement gives the investigator extremely powerful control over a variety of rates and patterns of responding.

The responses studied in operant conditioning experiments are usually easy to record automatically, and the need for arranging precisely-scheduled reinforcement contingencies leads the experimenter to seek reliable programming equipment. It is sometimes wrongly assumed that the use of automatic electrical programming and recording equipment is a distinguishing characteristic of operant conditioning experiments. This type of equipment has long been used in psychophysics laboratories, and in recent years techniques that formerly required the manual labor of the investigator (or an energetic graduate assistant) have been modified to permit automatic control of experiments. For example, several investigators have developed automated versions of the Wisconsin General Test Apparatus (see Chapter 1 by Meyer *et al.*). As experimental psychology progresses, an increasing number of investigators will be using automatic equipment.

Operant conditioning techniques have been used to investigate a wide range of problems. It is sometimes wrongly assumed that study of a simple, readily-repeatable response is necessarily limited because it is only one of a wide variety of possible measures. The notion that it is necessary to use a variety of measures in behavioral experiments has plausibility and intuitive appeal. However, the study of simple, easily-recorded responses has characteristically facilitated the advancement of knowledge in the biological sciences. In autonomic pharmacology, for example, studies of the contraction of the nictitating membrane of the cat, a simple and seemingly restricted response, have made a major contribution to the understanding of the modes of action of the autonomic nervous system. The nictitating-membrane response is important to the experimental pharmacologist because it is differentially sensitive to a number of relevant variables. A simple operant response, such as lever-pressing, is important to the experimental psychologist for similar reasons. The present chapter will provide ample evidence that rate and pattern of occurrence of a simple operant response are differentially sensitive to a variety of important variables.

In recent years, operant conditioning studies have emphasized a new approach to comparative psychology. The traditional approach in comparative psychology has been to study the behavior of different species in nearly identical experimental situations. Operant conditioning studies have attempted to vary experimental situations until the behavior of diverse species can be made as similar as possible. This new approach not only provides information about the particular stimuli and responses that are most suitable for developing a particular pattern of behavior in a species, but it also provides a common starting point for comparing the effects of a given variable upon diverse species. Operant conditioning studies of the behavior of different species are giving psychologists a new perspective on the comparative behavior of animals. This chapter will discuss two areas in which operant conditioning experiments have contributed to our understanding of the behavior of nonhuman primates.

II. BEHAVIOR CONTROLLED BY AVERSIVE STIMULI

Psychologists have long been interested in studying the ways in which the behavior of nonhuman primates can be affected by aversive stimuli. In technical terms, an aversive stimulus is a stimulus that an organism will work to escape from or to avoid; for example, a severe electric shock is an aversive stimulus for most organisms. Operant conditioning experiments emphasize the study of the effects of aversive stimuli on the rate or pattern of occurrence of responses that are related in different ways to the presentation of these stimuli. This section will discuss experimental studies of behavior controlled by scheduled presentations of aversive stimuli.

A. Avoidance Behavior

Primates are excellent subjects for studies of avoidance behavior. On avoidance schedules, responses prevent the occurrence of aversive stimuli. The type of avoidance schedule that will be considered in this section, continuous avoidance, was first described by Sidman (1953a). Brief shocks are delivered to the subject at regular shock-shock intervals (S-S) unless the subject responds; each response postpones the shock for a specified response-shock interval (R-S). If the subject responds frequently enough, no shocks are delivered. Rate of responding provides a convenient quantitative indication of the strength of avoidance behavior.

The investigator who has trained both rats and primates on continuous-avoidance schedules cannot fail to be impressed with the differences in the performances that are established. Once responding has developed, the rat typically receives a large number of shocks, especially at the start of each session, but the primate seldom receives a shock because it responds at a very high rate. Although differences in the avoidance behavior of rats and monkeys have not been directly studied, it appears that rats are more sensitive to changes in S-S and R-S intervals, probably because their avoidance behavior extinguishes more readily than that of monkeys. The parametric studies of S-S and R-S intervals that Sidman (1953b) conducted with rats should be replicated with primates.

An avoidance-conditioning apparatus for chimpanzees (*Pan*) or other large primates is shown in Fig. 1. The animal is restrained by a web harness, and a metal plate around the animal's waist prevents it from reaching its feet. Shock can be delivered through electrodes attached to the animal's feet. The avoidance response is pressing the lever on the panel in front of the animal. In one study (Clark, 1961), chimpanzees were trained on the following continuous-avoidance procedure. Brief electric shocks were delivered at regular 3-second S-S intervals unless

FIG. 1. Chimpanzee in restraining chair. The animal postpones shocks by pressing the lever. (From Clark, 1961.)

the animal responded; each response postponed the shock for a 20-second R-S interval. Representative results showing the development of avoidance performance for one chimpanzee are presented in Fig. 2. In the initial sessions, brief periods of responding occurred after some shocks, but numerous shocks were delivered (Records A and B). In the next session (Record C) relatively prolonged periods of negatively-accelerated responding occurred after each shock. The final performance (Records D and E) was characterized by relatively stable rates of responding that were high enough to avoid almost all shocks. These results are consistent with those obtained by Sidman (1958) with rhesus monkeys (*Macaca mulatta*) and by Kelleher and Cook (1959) with squirrel monkeys (*Saimiri sciureus*). Ferster (1958a) has shown that chimpanzees will also respond at a high rate on a continuous-avoidance schedule to postpone the start of a period of time out from a schedule of positive reinforcement.

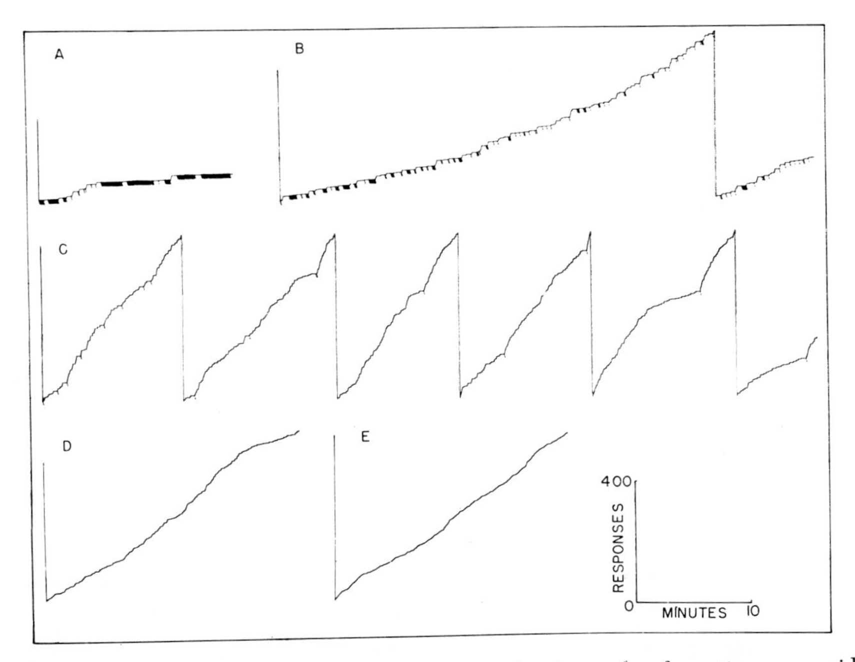

FIG. 2. Cumulative response records showing development of continuous avoidance performance in a chimpanzee. Each lever-pressing response moved the recording pen up the ordinate; the pen reset to the bottom of the record after about 550 responses had accumulated. Pips indicate the delivery of brief electric shocks. (From Clark, 1961.)

Continuous-avoidance behavior can be brought under the control of exteroceptive stimuli in a multiple schedule. For example, Hearst *et al.* (1960) trained a rhesus monkey on a discrimination procedure in which 5-minute periods of a clicking sound alternated with 5-minute periods of silence throughout each 3-hour experimental session. During the periods of clicking sound, a continuous-avoidance schedule was in effect with 5-second S-S and R-S intervals. During the periods of silence, shocks were not delivered and responses had no specified consequences. The cumulative response records in Fig. 3 indicate that high response rates (averaging about 120 responses per minute) in the presence of the clicking sound alternated with low response rates (averaging about 3 responses per minute) in its absence.

Although the results shown in Fig. 3 demonstrate that continuous avoidance can be controlled by discriminative stimuli, several investigators have noted that it is difficult to establish this type of control (Appel, 1960; Hearst, 1960, 1962; Sidman, 1961). Hearst (1960, 1962) directly compared generalization gradients based on training on a schedule of positive reinforcement with those based on continuous-avoidance training. In one experiment, rhesus monkeys were trained to press a lever on a

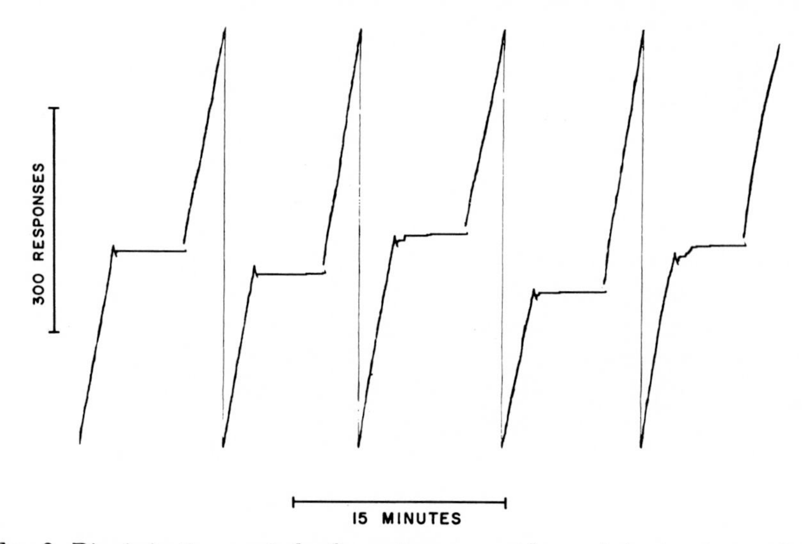

FIG. 3. Discriminative control of continuous avoidance behavior in a rhesus monkey. (From Hearst *et al.*, 1960.)

continuous-avoidance schedule (S-S and R-S intervals were 10 seconds), and concurrently to pull a chain on a 2-minute variable-interval (VI-2) schedule of food reinforcement. The light in the experimental chamber was at maximum intensity throughout this training. During generalization testing, the light was at various lower intensities presented in an irregular series, and neither shock nor food was delivered. As the intensity of the light was decreased, the relative frequency of food-maintained responses decreased rapidly (steep generalization gradient), but the relative frequency of shock-maintained responses remained high even at the lowest intensity (flat gradient).

In a second experiment (Hearst, 1962), the same rhesus monkeys received discrimination training at the two extreme light intensities before generalization testing; that is, the concurrent avoidance and VI schedules were in effect at one extreme light intensity while experimental extinction was in effect at the other extreme. Discrimination training was continued until response rates were low in the presence of the stimulus correlated with experimental extinction. Both generalization gradients were steeper after discrimination training (see also Sidman, 1961), but the avoidance gradient remained flatter than the VI gradient.

These experiments show that characteristics of stimulus generalization gradients can depend critically on the reinforcement schedule used in training. The use of concurrent schedules, involving different responses in the presence of the same stimuli, shows that the differences between the two types of generalization gradient are not produced by stimulus preferences or by differences in the discriminability of stimuli. These data provide further support for Blough's (1961) suggestion that there is no meaningful relation between discrimination (j.n.d.) functions and those regions of a stimulus dimension where generalization occurs most readily. But these experiments do not indicate which of the many attributes of schedule-controlled behavior determine the differences in generalization gradients. For example, the results may reflect differences in initial response rates, in motivation, or in extinction functions.

The notion that characteristics of stimulus generalization gradients are a function of motivational variables is appealing because it might provide a plausible account of seemingly irrational behavior in fearful or anxious humans. In his account of psychoanalytic displacement, Miller (1948) cited evidence for the assumption that generalization gradients are steeper after training with aversive stimulation than after training with positive reinforcement. The experiments by Hearst could be interpreted as supporting the opposite assumption. The discrepancy probably stems from the use of different types of schedules of aversive stimulation and of positive reinforcement. In the experiments cited by Miller, aversive stimuli were delivered either independently of responses or while the subjects were eating, so the omission of shocks during generalization testing was an abrupt change, which would facilitate extinction. On the other hand, Hearst used a continuous-avoidance schedule, and the omission of shocks during generalization testing was a minimal change which would tend to slow extinction. Little is known about the interaction between resistance to extinction and the characteristics of stimulus generalization.

When performance on a schedule of positive reinforcement differs from performance on a schedule of aversive stimulation, it has usually been assumed that the difference can be attributed to differences in motivation even though the schedules are not formally comparable. Although this assumption seems plausible, it is becoming increasingly apparent that schedules can be as important as motivational variables. In order to assess the role of motivational variables in determining generalization gradients, it will be necessary to develop schedules of aversive stimulation that are formally comparable to schedules of positive reinforcement. Such schedules are now being developed, as described in the next section.

B. Escape Behavior

On escape schedules, responses terminate aversive stimuli. Squirrel monkeys will respond at high rates on an escape procedure in which responses intermittently terminate an exteroceptive stimulus in the presence of which unavoidable electric shocks are presented (Azrin *et al.*, 1962). In the first phase of the procedure, the monkeys received brief unavoidable foot shocks at irregular intervals of time, averaging 2 minutes, in the presence of a visual stimulus. When the stimulus was absent (safe period), shocks were not presented. Delivering shocks in the presence of the visual stimulus established it as an aversive stimulus. In the second phase, the monkey could terminate the stimulus and produce a 2-minute safe period by pressing a lever. In the third phase, the number of responses necessary to terminate the stimulus was gradually increased. This schedule of stimulus termination is formally comparable to a fixed-ratio (FR) schedule of food reinforcement. When the number of responses required to produce the safe period was 150 or less, the monkeys responded at a high rate in the presence of the stimulus. At response requirements of 250 or 350, a pause after each safe period was followed by a high response rate that was maintained until the stimulus terminated. Shocks delivered in the presence of the stimulus occurred while the monkey was responding rapidly, but high response rates were maintained. Under this escape procedure, the scheduled relation between responses and termination of the stimulus is apparently more important than the unscheduled relation between responses and shocks.

In a subsequent investigation, Azrin *et al.* (1963) studied the effects of varying shock frequency in the presence of the stimulus. The FR response requirement was 25 for some squirrel monkeys and 50 for others. The general training procedure was similar to that described for the previous study. Figure 4 shows the effects of different shock frequencies on performance of a monkey that was required to make 25 responses to terminate the stimulus for 60 seconds. The lower part of Fig. 4 shows that the monkey made about 8,000 responses in each 6-hour session at shock frequencies ranging from 1 to 30 per hour. About 5,000 responses occurred when the shock frequency was once in 2 hours; however, the number of responses was drastically reduced when the shock frequency was once in 6 hours. When no shocks were delivered, responding ceased. The cumulative response records in the upper part of Fig. 4 show that decreasing the shock frequency below two per hour produced pauses after each safe period; however, when responding started, it occurred at a high rate until the stimulus terminated.

Azrin *et al.* (1963) also studied the effects of varying the duration of the safe period. Figure 5 shows the effects of different durations of the safe period on performance of one squirrel monkey on a FR-50 schedule

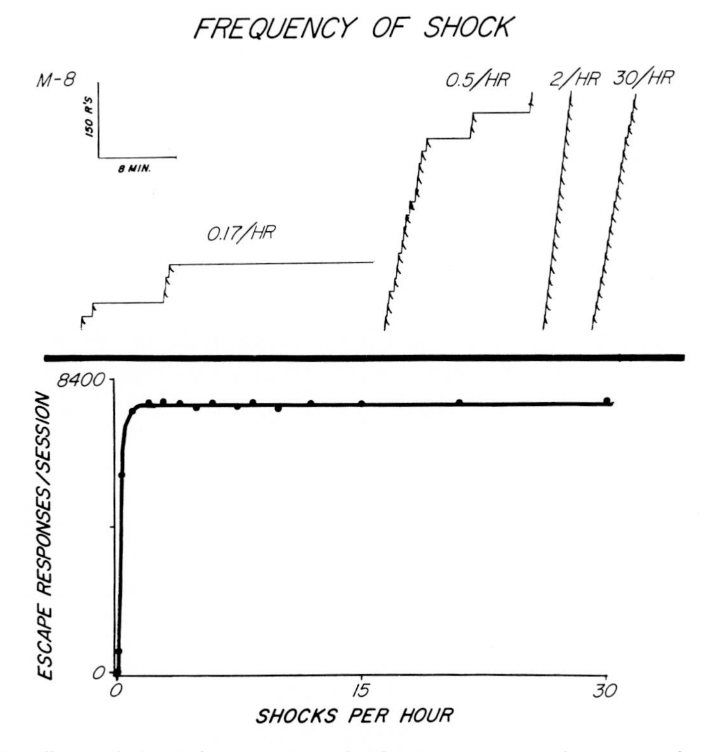

FIG. 4. Effects of shock frequency on fixed-ratio escape performance of a squirrel monkey. Bottom: total number of responses per 6-hour session as a function of shock frequency. Top: representative cumulative response records obtained at different shock frequencies. Pips indicate 60-second safe periods in which the recorder did not run. (From Azrin *et al.*, 1963.)

of escape with the shock frequency at six per hour. The lower part of Fig. 5 shows that response rates in the presence of the stimulus were above 180 per minute at safe-period durations ranging from 10 seconds to 240 seconds; response rates decreased markedly at durations of less than 10 seconds. The cumulative records in the upper part of Fig. 5 show that decreasing the duration below 10 seconds produced pauses following each safe period; these pauses were terminated by abrupt changes to high response rates that were maintained until the stimulus terminated.

The procedure developed by Azrin *et al.* provides an excellent technique for investigating other schedules of intermittent termination of a conditioned aversive stimulus, and for comparing them with schedules of positive reinforcement. On the continuous-avoidance schedules described in Section II, A, monkeys usually responded at a rate that was high enough to keep the shock frequency below one per hour; however,

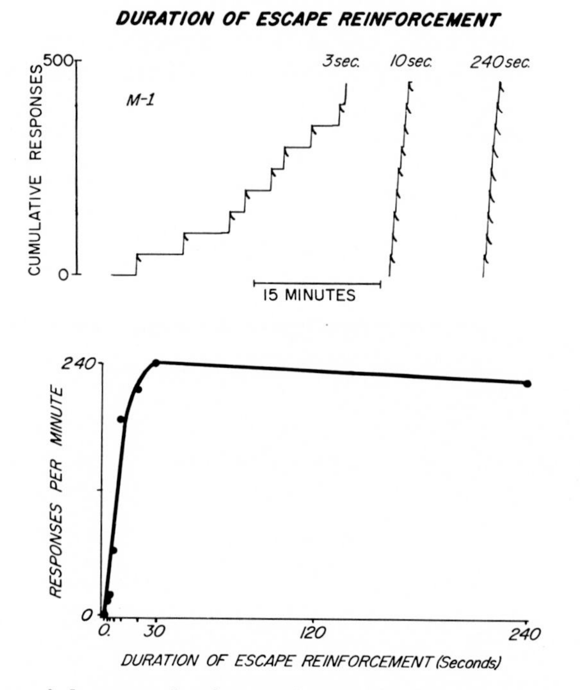

FIG. 5. Effects of duration of safe period (escape reinforcement) on escape performance of a squirrel monkey. Bottom: response rate as a function of duration of safe period. Top: representative cumulative response records obtained at different durations of escape. Pips indicate safe periods in which the recorder did not run. (From Azrin *et al.*, 1963.)

if a monkey did not respond when the S-S interval was 30 seconds, for example, it would receive 120 shocks per hour. On the FR-escape procedure described in this section, high response rates were maintained over long time periods, even when an animal would receive only one shock per hour if it did not respond. Results with both procedures show that relatively infrequent shocks can maintain continuously high response rates.

When responses did not produce the safe period (experimental extinction), high response rates alternated with progressively lengthening pauses. When extinction was continued over several sessions, high response rates prevailed in the early part of each extinction session. Azrin *et al.* (1963) concluded that performances on FR schedules of escape have most of the characteristics of performances on FR schedules of food reinforcement. In experimental extinction after continuous-avoidance training, responding is initially well maintained and there are few occasions on which shock would have occurred if extinction were not in effect. After training under an escape schedule, however, responding in the initial phases of extinction produces many occasions on which safe periods would have occurred if extinction were not in effect. Generalization gradients determined after escape training on a VI schedule might be comparable to generalization gradients determined after training on a VI schedule of food reinforcement. The establishment of performances on various schedules of escape should greatly facilitate experimental analyses of motivational effects (cf. Kelleher & Morse, 1964).

C. Effects of Unavoidable Shock

Estes and Skinner (1941) demonstrated that a stimulus that regularly precedes an unavoidable shock (preshock stimulus) can suppress responding that is maintained by a schedule of food reinforcement. This basic finding has been confirmed with several species on several schedules of food reinforcement. Figure 6 shows the effects of presenting a 5-minute preshock stimulus to a rhesus monkey working on a VI schedule of food reinforcement. The cumulative response records show that responding is almost completely suppressed in the presence of the preshock stimulus even though the VI schedule is still in effect.

Sidman *et al.* (1957) first observed that under some conditions a preshock stimulus would increase rather than suppress responding of rhesus monkeys. The monkeys were trained to respond on a continuous-avoidance schedule (S-S and R-S intervals were 20 seconds) in the presence of Stimulus A. When performance had developed on this schedule, a preshock stimulus (Stimulus B) was superimposed on the continuous-avoidance schedule. In each session, 5-minute periods of Stimulus A alternated with 5-minute periods of Stimulus B. The continuous-avoidance schedule remained in effect during presentation of each of these stimuli, but Stimulus B terminated with a brief unavoidable shock. Introducing the preshock stimulus and the unavoidable shock produced almost a three-fold increase in rate of responding. Response rates later decreased, but they decreased more slowly in the presence of the preshock stimulus. Thus, at some phase of the experiment, response rates were higher in the presence of the preshock stimulus than in its absence. When the avoidance response was extinguished by the omission of avoidable shocks, the response rate again decreased more slowly in the presence of the preshock stimulus than in its absence. Higher response rates in the presence of the preshock stimulus persisted for as long as 300 hours of avoidance extinction. Subsequent studies (Appel, 1960; Hearst, 1962; Sidman, 1958; Sidman *et al.*, 1962) have shown that the continuous-avoidance response rates of rhesus monkeys increase markedly whenever shocks are delivered unavoidably.

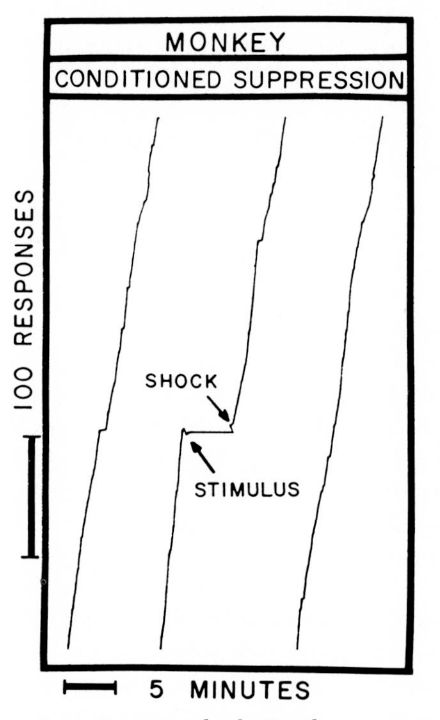

FIG. 6. Effects of superimposing a preshock stimulus on responding maintained by a VI schedule of food reinforcement in a rhesus monkey. (From Sidman, 1960.)

Following training on continuous avoidance, unavoidable shocks can increase responding even in the presence of a stimulus in which shocks have not previously occurred. Appel (1960) trained two rhesus monkeys on a discrimination procedure in which 5-minute periods of Stimulus A alternated with 5-minute periods of Stimulus B. In the presence of Stimulus A, S-S and R-S intervals of 20 seconds were in effect. In the presence of Stimulus B, shocks were not delivered and responses had no specified consequences. The monkeys responded at a high rate in the presence of Stimulus A and at a low rate in the presence of Stimulus B (see Fig. 3). When unavoidable shocks were then delivered at irregular intervals (averaging 3 minutes) in the presence of Stimulus B, response rates increased markedly in the presence of this stimulus. Hearst (1962) has independently confirmed this finding. A discrimination developed with continuous-avoidance procedures can be eliminated by the introduction of unavoidable shocks.

A critical experiment by Herrnstein and Sidman (1958) first demonstrated that the suppressant effects of a preshock stimulus can actually be reversed by interpolated avoidance-training sessions. The subjects were rhesus monkeys. In the first phase of the experiment, performance on a 1-minute variable-interval (VI-1) schedule of food reinforcement was suppressed by introducing a clicking sound that preceded an unavoidable shock. In the second phase, the clicking stimulus and unavoidable shock were absent, and the monkeys responded on a continuous-avoidance schedule (S-S and R-S intervals were 20 seconds). In the third phase, the monkeys again responded on the VI-1 schedule of food reinforcement. In the fourth phase, the clicking stimulus followed by unavoidable shock was again superimposed on the VI-1 schedule; however, it no longer suppressed responding. In fact, the monkeys responded at a higher rate in the presence of the preshock stimulus than they did in its absence. When avoidance responding was extinguished before the fourth phase, response rates were lower in the presence of the preshock stimulus than in its absence. This study showed that the effects of an aversive stimulus could be completely dependent on the experimental history of the monkey.

A recent series of experiments with squirrel monkeys investigated the extent to which a history of avoidance conditioning could influence the effects of a preshock stimulus (Kelleher *et al.*, 1963). In the first phase of this investigation, 10-minute periods of a VI-1.5 schedule of food reinforcement in the presence of a white light alternated with 10-minute periods of a shock-avoidance schedule (S-S and R-S intervals of 30 seconds) in the presence of a yellow light. A 2-minute blue light that terminated with an unavoidable shock was then superimposed on the VI-1.5 schedule. By the twentieth 4-hour session on this procedure, response rates were highest in the presence of the preshock stimulus. Record A of Fig. 7 shows the performance of one monkey after about 220 sessions on this procedure. Even with prolonged exposure to the schedule, the highest response rates continued to occur in the presence of the preshock stimulus. As shown in Record B of Fig. 7, these high response rates persisted even when responses were no longer reinforced by food in the presence of the preshock stimulus.

Next, extinction was instituted by omission of both food reinforcements and avoidable shocks; however, unavoidable shocks still occurred at the termination of the blue light. Under these conditions response rates decreased to near zero in the presence of the white and yellow lights, but remained high in the presence of the blue light. Again these findings indicate that the effects of an aversive stimulus can be markedly altered by previous training. They also indicate that the effects of continuous-avoidance training can persist even when the avoidance responding itself has been extinguished. In subsequent experiments with the same squirrel monkeys, behavior was maintained by the unavoidable shock alone. Ten-minute periods of the white light alternated with 2-minute periods of the

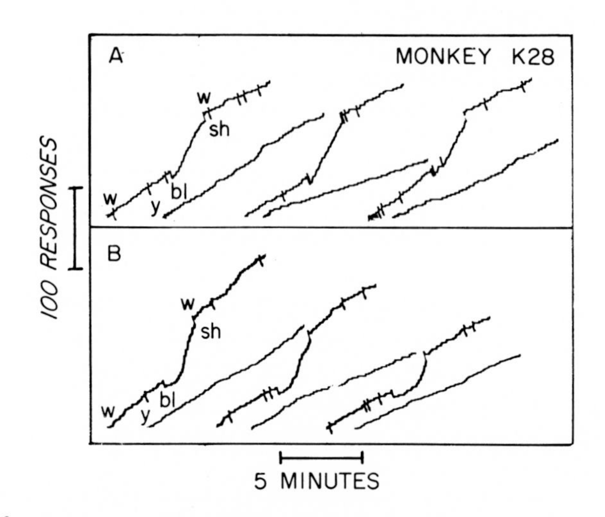

FIG. 7. Performance of a squirrel monkey on a schedule including a continuous-avoidance component and a VI component with a preshock stimulus. The 10-minute segments of records have been displaced along the abscissa. The letters adjacent to the first segment of each record indicate the order of stimuli. The first segment in Record A starts with VI 1.5 with the white (w) light on; the offset portion shows VI 1.5 with the blue (bl) light on terminating with an unavoidable shock (sh); this is followed by VI 1.5 with the white light on. The pips when the white or blue lights were on indicate food reinforcements; the recorder did not run during reinforcement cycles. The second segment shows continuous avoidance with the yellow (y) light on. The sequence is identical in Record B, but food reinforcement could not occur when the blue light was on. (From Kelleher *et al.*, 1963.)

blue preshock-stimulus light. Note that under these conditions lever-pressing had no explicitly programmed consequences. Record A of Fig. 8 shows the performance of a squirrel monkey on this procedure. Positively accelerated responding occurred during each presentation of the preshock stimulus. When unavoidable shocks were omitted, as shown in Record B of Fig. 8, the monkey stopped responding. When unavoidable shocks were reinstated, as shown in Records C and D of Fig. 8, the monkey responded again after several shocks were delivered. The unavoidable shock was the essential condition maintaining responding in the presence of the preshock stimulus.

In the final experiment by Kelleher *et al.*, the preshock stimulus was present throughout each experimental session, and brief unavoidable shocks were delivered at 10-minute intervals. The performances that developed on this procedure are shown in Fig. 9. These cumulative records show some instances of positively accelerated responding between shocks; however, the acceleration sometimes became negative before

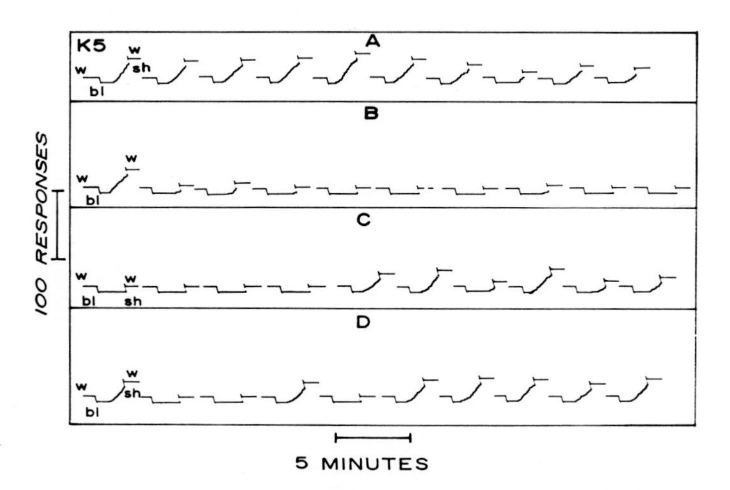

FIG. 8. Performance of a squirrel monkey in the presence of the blue (bl) light as a function of removing and then reinstating the unavoidable shock. (A) Unavoidable shock terminated each presentation of the blue light; (B) unavoidable shocks were not delivered; (C) first session in which shocks were reinstated; (D) final performance with shocks terminating each presentation of the blue light. (From Kelleher *et al.*, 1963.)

the shock was delivered. The substantial levels of behavior shown in Fig. 9 were maintained only by the intermittent delivery of unavoidable electric shocks.

Why was responding maintained by unavoidable shocks? Some investigators have interpreted increased responding in the presence of a preshock stimulus as "superstitious" avoidance responding (Herrnstein & Sidman, 1958; Sidman, 1958; Sidman *et al.*, 1957). That is, on a continuous-avoidance schedule each response postpones shock, but whenever the subject pauses longer than the R-S interval, a shock is delivered. Although unavoidable shocks are scheduled to occur independently of responses, they will actually occur at variable intervals of time after a response. Although the variable temporal relation between responses and unavoidable shocks has not been programmed by the experimenter, the subject might continue to be controlled by the time intervals between responses and shocks. Sidman and Boren (1957) have demonstrated that high levels of responding can be maintained in rats by programming variable R-S intervals.

It is reasonable to try to interpret continued responding in the presence of a preshock stimulus as an instance of superstitious avoidance conditioning. Some experiments have provided ideal conditions for the development of superstitious avoidance responding (for example, Sidman *et al.*, 1957); however, conditions are more unusual in the others.

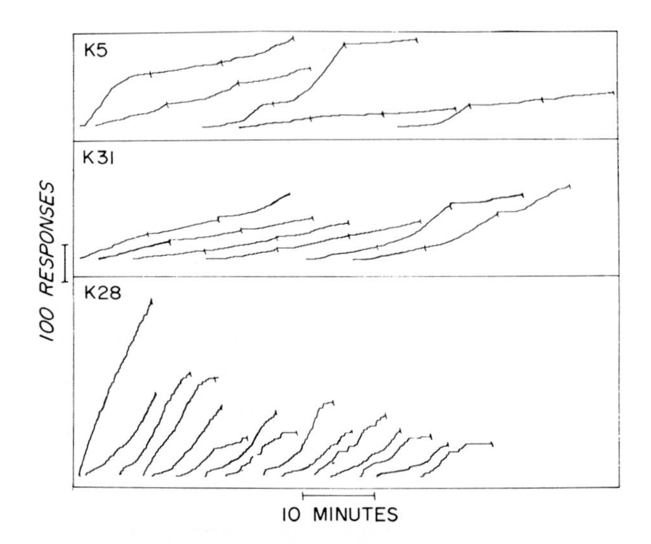

FIG. 9. Performance of three squirrel monkeys with blue light on continuously. The pips on the cumulative records indicate the delivery of unavoidable shocks at regular 10-minute intervals. The records of monkeys K5 and K31 have been broken into 30-minute segments and displaced along the abscissa. The records of monkey K28 are in 10-minute segments. (From Kelleher *et al.*, 1963.)

For example, Herrnstein and Sidman (1958) noted that, when the monkey was returned to the VI schedule after avoidance training, the first presentation of the preshock stimulus increased response rates. Since the preshock stimulus had not been present during avoidance training, and an unavoidable shock had not yet occurred following responding in the presence of the preshock stimulus, this initial increase in responding during presentation of the preshock stimulus is not easily explained in terms of accidental correlations. It is also difficult to use the notion of superstition to explain results showing that the responding will persist in the presence of a preshock stimulus even after avoidance responding has been extinguished (Kelleher *et al.*, 1963; Waller & Waller, 1963).

Responding maintained by unavoidable shock has been demonstrated in chimpanzees (Belleville *et al.*, 1963), rhesus monkeys (Herrnstein & Sidman, 1958), squirrel monkeys (Kelleher *et al.*, 1963), and dogs (Waller & Waller, 1963). This behavioral phenomenon could be called "abnormal," but, as Sidman (1960) has indicated, we must specify what is meant by the term. Because this "abnormal" behavior typically develops under the experimental conditions described above, it is the experimental procedure rather than the behavior that is abnormal. To the extent that we can specify the essential aspects of these experimental conditions, we can understand and control the development of the abnormal behavior.

Many questions remain to be answered by future studies of these surprising behavioral effects. Are the effects of a history of continuous-avoidance training irreversible? Can the effect be obtained after training on other schedules of aversive stimulation, such as the escape schedule described in Section II, B? Can animals be trained under conditions in which they will respond to produce an aversive stimulus? An experimental program designed to answer such questions necessarily focuses attention on unusual ways in which potent behavioral variables can interact to determine behavior. This approach to the abnormal behavior of nonhuman primates should add much to our understanding of the abnormal behavior of human primates.

III. DISCRIMINATIVE BEHAVIOR

Comparative psychologists have devoted a substantial proportion of their research to studies of learned behavior of nonhuman primates. It is natural that the human primate is particularly interested in the intellectual capabilities of his closest relatives on the phylogenetic scale. It is well to remember, however, that in the comparative studies of learned behaviors, the investigator must strive to maintain his objectivity. It is far more tempting to infer intelligence when observing performance of a monkey or a chimpanzee than when observing similar performance of a rat or a pigeon. Fortunately, objective techniques are available for studying a wide range of learned behaviors in animals.

The experimental techniques most frequently used for the study of learned behavior of nonhuman primates involve discriminative behavior. That is, the investigator studies the probability of a response as a function of changes in the animal's environment. For example, a reinforcing stimulus may be presented if a monkey responds at a high rate when a triangular stimulus is present but does not respond when a circular stimulus is present. This section will describe operant conditioning studies of discriminative behavior. These studies will provide the empirical basis for a general discussion of problems in the comparative analysis of learned behavior.

A. Multiple Schedules of Reinforcement

Different kinds of schedules of reinforcement can be used to establish a wide variety of rates and patterns of responding. In a multiple schedule, several different types of schedule-controlled behavior are developed by putting each schedule of reinforcement in effect in the presence of a different exteroceptive stimulus (Ferster & Skinner, 1957). The stimuli and their corresponding schedules may be presented in a regular or an irregular sequence.

Multiple schedules permit the investigator to study several different kinds of behavior in a single animal within a short period of time. The degree of complexity that can be arranged in a multiple schedule is limited only by the available number of response topographies, types of reinforcer, schedule conditions, stimulus conditions, and combinations and permutations of these four variables. Multiple schedules that include schedules of aversive stimuli and schedules of food reinforcement have already been described in Section II, C. The present section will discuss three multiple schedules that illustrate different levels of complexity.

1. Simple Stimulus Conditions

A multiple schedule that involves a regular sequence of three stimuli has been used to study the behavior of squirrel monkeys (Cook & Kelleher, 1962). In the presence of Stimulus A, a 10-minute fixed-interval (FI-10) schedule was in effect; that is, the first response occurring after 10 minutes was reinforced by food. In the presence of Stimulus B, a 30-response fixed-ratio (FR-30) schedule was in effect; that is, the 30th response was reinforced by food. In the presence of Stimulus C, a 2.5-minute time-out (TO-2.5) period was in effect; that is, responses were not reinforced.

The cumulative response records labeled "control" in the upper part of Fig. 10 show a typical performance of a squirrel monkey on this three-component multiple schedule. In the FI-10 components (as at *a*), the monkey paused for several minutes, and then responding increased until reinforcement occurred. In TO-2.5 components (as at *b* and *d*), responding seldom occurred. In FR-30 components (as at *c*), the monkey responded at a high rate.

For studying variables that might affect discriminative behavior, this type of multiple schedule has several advantages over the type of discriminative procedure that indicates only whether or not a response occurred when a stimulus was presented. For example, the middle and lower records in Fig. 10 show typical effects of oral doses of meprobamate (Miltown®). After 50 mg/kg of meprobamate, average rate of responding in the FI-10 component was increased markedly. However, the records show that performance in the FI-10 component was still characterized by a pause followed by positively accelerated responding, and performances in the TO-2.5 and FR-30 components were un-

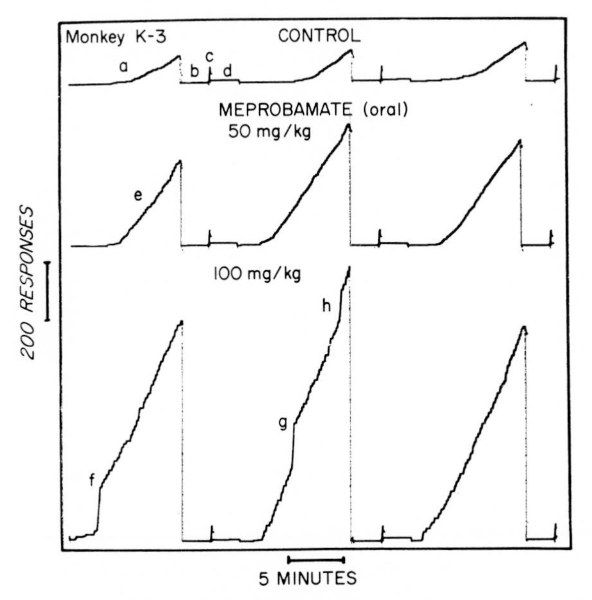

FIG. 10. Effects of meprobamate on a squirrel monkey's behavior maintained by a multiple schedule of food reinforcement including FI-10 (*a*), FR-30 (*c*), and TO-2.5 (*b* and *d*) components. The recording pen reset to the bottom of the record when reinforcement occurred. (From Cook & Kelleher, 1962.)

affected. This dose of meprobamate had a large effect on the monkey's behavior, but differential response patterns still occurred in the presence of each of the three stimuli. After 100 mg/kg of meprobamate, rate of responding in the FI-10 component was further increased; however, the pattern of responding was disrupted. Bursts of responding (shown at *f*, *g*, and *h*), which occurred at a rate that was comparable to the response rate in the FR-30 component, indicate that stimulus control in the FI and FR components had been disrupted. Performance in the TO-2.5 component was still differentiated from the other two components. It should be added that 100 mg/kg of meprobamate produces observable quieting and ataxia in squirrel monkeys. This high dose of meprobamate disrupted one aspect of the discriminative control but did not affect others. The use of the multiple schedule shows that the effect of meprobamate is selective.

Performances of rhesus monkeys (Brady, 1958; Skinner, 1956) and chimpanzees (Kelleher, 1957) are comparable to those of squirrel monkeys on the type of schedule shown in Fig. 10. Indeed, the performances of nonprimates, such as rats and pigeons, are also comparable to the performances of primates on this type of schedule (Skinner, 1956). At this point, the comparative psychologist might assume that multiple-schedule performance is of little interest because it does not even separate primates from nonprimates. Such an assumption would be misleading. Establishing comparable discriminative behaviors in squirrel monkeys and in pigeons, for example, is an essential starting point for determining the effects of other variables on the performances of each of these different species. An excellent example is again provided by drug effects. The oral administration of a range of doses of chlorpromazine (Thorazine®) to monkeys and pigeons on multiple FI-10 FR-30 schedules has qualitatively different effects on the behaviors of the primate and the nonprimate. In the pigeon, doses of 1 to 10 mg/kg have little effect on FR-30 or TO-2.5 performance, but produce marked changes in the pattern of responses on FI 10; the characteristic curvature of the cumulative records within each FI-10 component is abolished because the bird responds at a relatively constant rate throughout (Fry *et al.*, 1960). In the squirrel monkey, however, doses of 0.3 mg/kg and higher produce a dose-dependent decrease in responding in both FI-10 and FR-30 components of the multiple schedule (Cook & Kelleher, 1962). Such results indicate either that the effects of chlorpromazine on a single behavioral process depend on the species used or that chlorpromazine acts on different behavioral processes which underlie the seemingly identical performances of the primate and nonprimate. The former alternative seems more probable. While behavioral processes are often quite general, it is well known that there are marked species differences in the effects of drugs. Investigations of the effects of different variables on behaviors that are comparable in the primate and the nonprimate should be an important part of the comparative psychology of the future.

2. Complex Stimulus Conditions

By varying the stimulus conditions that are correlated with each component of a multiple schedule, extremely complex discriminative behaviors can be studied. One type of multiple schedule has been used to study concept formation by chimpanzees (Kelleher, 1958). The stimuli were made up of illuminated square windows in a 3-by-3 matrix. Each of the 13 positive stimulus patterns had some common element that was not present in any of the 13 negative patterns. In the presence of positive stimulus patterns, a 100-response variable-ratio (VR-100) schedule of food reinforcement was in effect; that is, every 100th response, on the average, was reinforced by food. Positive stimulus patterns terminated at reinforcement. In the presence of negative stimulus patterns, responses were not reinforced. Negative stimulus patterns terminated after 1 minute without a response.

Frame A of Fig. 11 shows representative positive and negative stimuli, as well as typical performances that developed in the presence of each

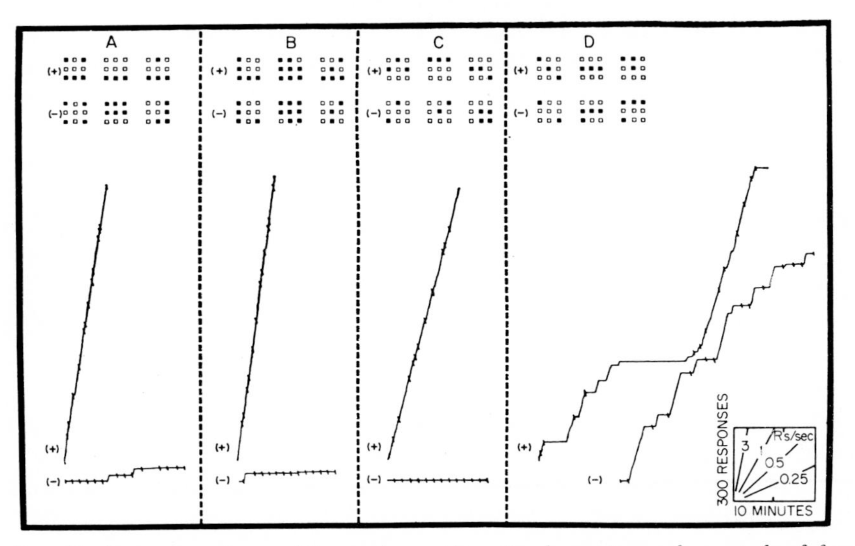

FIG. 11. Representative positive and negative stimulus patterns from each of four sequences are shown in the upper portion of each frame. Cumulative response records from each of the four sequences are shown in the lower portion. Responses during presentation of positive and negative stimuli were cumulated separately. Pips indicate where stimulus presentations terminated. The recorders did not run during the 30-second time-out periods between stimulus presentations. (From Kelleher, 1958.)

type of stimulus. The dark squares in Fig. 11 correspond to illuminated windows in the matrix. The common element of the positive stimuli was the illumination of the bottom row of windows. During presentations of positive stimuli, the chimpanzees responded at the high, stable rate that usually develops on VR-100 schedules. During presentations of negative stimuli, the animals seldom responded. Changing the sequence of presentation of the stimuli did not affect performance. Next, six of the positive and six of the negative stimuli were changed without changing the common element of the positive stimuli. These changes are exemplified in the upper part of Frame B of Fig. 11. The results presented in the lower part of Frame B show that these changes did not disrupt performance. The different performances in the presence of positive and negative stimuli were based on the common element of the positive stimuli.

In a second experiment with these chimpanzees, the common element of positive stimuli was the illumination of three windows. In negative stimuli, either two or four windows were illuminated. Frame C of Fig. 11 shows representative positive and negative stimuli and typical cumulative response records after performance had developed on this second concept problem. Again, changing the sequence of presentation of the stimuli did not affect the performance shown in Frame C. However, when six positive and six negative stimuli were changed without changing the common element of the positive stimuli, performance was markedly disrupted in the presence of each of the 12 new stimuli. Frame D of Fig. 11 shows representative stimulus changes and typical performance in the session in which the new stimuli were introduced. In this case, the discriminative performances shown in Frame C of Fig. 11 were not based on the common element of the positive stimuli.

The chimpanzees developed stable discriminative performances on each of the complex multiple schedules just described. Neither of these performances was disrupted by changing the sequence in which the stimuli were presented. Performances on the second schedule, however, were disrupted by changing specific stimulus patterns. Although the initial performances in each procedure were almost identical, the basic processes involved in these performances were qualitatively different. On the second schedule, responding must have been based upon specific stimulus patterns; on the first schedule, responding was based upon the common element or concept. With the procedure just described, concept formation was a function of the concept problem.

This study of concept formation by chimpanzees raises several important points. The particular procedure that is used is an important determinant of the type of behavior that develops. One approach to the comparative study of learned behavior is to develop the most complex performance possible in each species studied. For example, the chimpanzees in the concept-formation study could probably have been trained to respond to the common element of the second concept problem if the stimulus patterns were repeatedly changed. In fact, concept formation has been demonstrated in monkeys (e.g., Weinstein, 1945). By using appropriate procedures, the investigator can develop performances of monkeys or chimpanzees that are similar to performances of humans. In this case, the comparative psychologist is not likely to assume that the performance is of little interest because it does not separate human from nonhuman primates. As noted above, such an assumption is usually misleading, and it should not be acceptable in comparing nonprimates and primates any more than it is in comparing nonhuman and human primates. What is required is a careful functional analysis of the behavioral processes involved in the performance of each species.

The extent to which discriminative behavior becomes controlled by particular stimuli or sequences of stimuli rather than by abstract attributes of stimuli may be an important characteristic of the behavior of nonhuman primates. In the concept-formation procedure used with chimpanzees, humans would undoubtedly respond to the common element of the positive pattern. In fact, humans would probably consider it relatively difficult to memorize each of the specific patterns in the second experiment. However, chimpanzees apparently learned to respond appropriately in the presence of each of the 26 specific stimuli that were repeatedly presented.

Similar results could have occurred in earlier studies of complex discriminations. For example, Nissen (1951) trained a chimpanzee on an exceedingly complex conditional discrimination in which the appropriate response on any trial depended on the way in which five stimulus variables were combined. Nissen noted that the animal could have responded appropriately by learning each of the 16 specific stimulus patterns, but he concluded that this was improbable. The concept-formation experiments suggest that response to the 16 specific stimuli could have occurred with the conditional-discrimination procedure (cf. Harlow, 1951, p. 198). Future studies should attempt to specify the variables that determine whether an animal will respond to specific stimuli or to an abstract characteristic of a set of stimuli.

3. COMPLEX STIMULUS AND SCHEDULE CONDITIONS

Belleville *et al.* (1963) recently described a complex multiple schedule that was developed to study the behavior of chimpanzees during space flight. This multiple schedule included a wide range of behaviors that could be sampled in a single chimpanzee over a short period of time in a small experimental space. A chimpanzee ("Enos") performed on a slightly modified version of this schedule on 29 November 1961, during two orbits around the earth in a space capsule (Rohles *et al.*, 1963).

The chimpanzees received preliminary training in the apparatus shown in Fig. 12. Three levers were mounted just below three corresponding display windows. Several colors or forms could be presented in each display window. The apparatus also contained a "lip-lever" drinking tube, mounted just below another stimulus light, and a food-pellet dispenser. Final training was conducted in a restraining chair similar to the one shown in Fig. 1. The stimulus displays, levers, and food and water dispensers were mounted on a panel directly in front of the animal. Electric shocks could be delivered through electrodes on the animal's feet.

The multiple schedule included four component schedules of reinforcement. The techniques used to develop performance on each of the component schedules have been described in detail by Belleville *et al.*

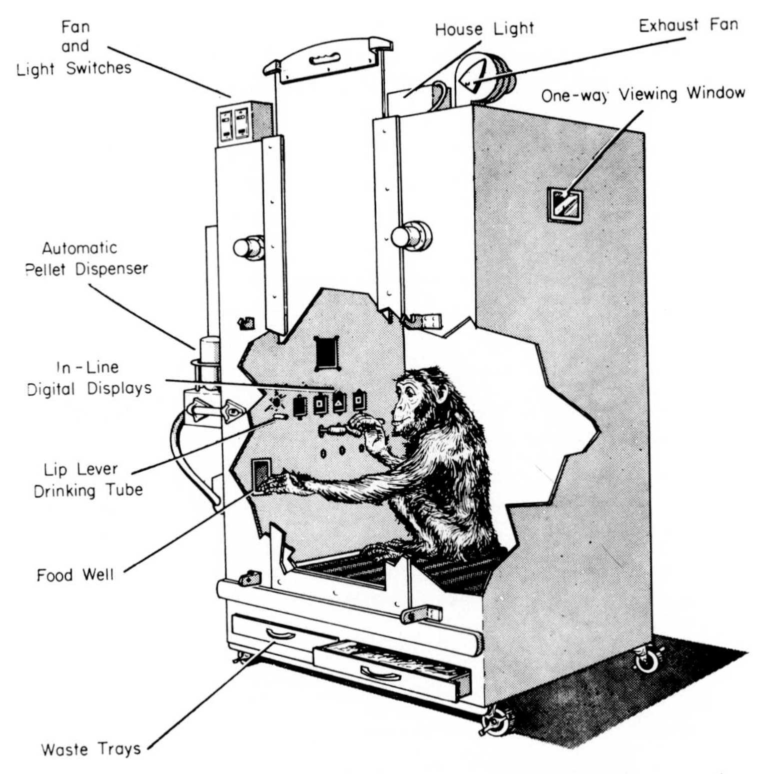

FIG. 12. Experimental chamber for chimpanzee. (After Rohles, 1961.)

(1963). A 2-minute time-out period followed each component. The schedules were presented in the following order:

(1) During the 10-minute period that Schedule A was in effect the right display window was red, and a continuous-avoidance schedule was in effect on the right lever (S-S interval was 2 seconds, and R-S interval was 10 seconds). At regular 2-minute intervals, the left display window became blue. If the chimpanzee did not press the left lever within 5 seconds after the blue stimulus appeared, a shock was delivered and the blue stimulus terminated; a response on the left lever terminated the blue stimulus and avoided the shock. Schedule A was referred to as "Avoidance."

(2) During the 10-minute period that Schedule B was in effect, the right display window was green; each response on the right lever that occurred after 10 seconds or more without a response turned on the stimulus light above the lip-lever drinking tube. In the presence of this stimulus light, pressure on the lip-lever drinking tube was reinforced by the delivery of water. Schedule B was referred to as "DRL."

(3) During the 10-minute period that Schedule C was in effect, the center display window was yellow, and every 50th response on the center lever was reinforced by the delivery of a food pellet. Schedule C was referred to as "FR 50."

(4) When Schedule D was in effect, a form stimulus was presented in each of the three display windows. Two of the forms were identical, but the third was different ("odd"). A response on the lever under the odd form was reinforced by the delivery of a food pellet, and the next set of three forms was presented. A response on the lever under either of the identical stimuli produced a 15-second time out, and then the same set of stimuli was presented again. The 18 sets that could be generated using circles, squares, and triangles were presented once in each cycle of the multiple schedule. Schedule D was referred to as "Oddity." The cumulative response records in Fig. 13 show the final performance of a chimpanzee on three consecutive cycles of the multiple schedule. Continuous-avoidance responses on the right lever occurred at a high rate. Responses on the left lever occurred quickly when the blue light was presented. No shocks were delivered during the three cycles shown. The DRL schedule controlled a low rate of responding on the right lever, and the chimpanzee obtained about two water reinforcements per minute. The animal responded at a stable, high rate on the FR-50 schedule, and more than 90% of its responses on Schedule D were responses to the odd stimulus.

The records in Fig. 13 show that the chimpanzee was almost perfectly controlled by the changing stimulus conditions, schedule conditions, and response conditions. The levels of proficiency attained on this complex multiple schedule led Belleville *et al.* (1963) to conclude that present procedures have not yet approached the limits of complexity for chimpanzees.

B. Intermittent Reinforcement of Discriminative Responses

In the multiple-schedule procedures discussed in Section III, A, different exteroceptive stimuli were presented successively, and resulting differences in rates and patterns of responding in the presence of each stimulus were analyzed. In many studies of discriminative behavior, however, the investigator wants frequent samplings of the animal's relative tendency to respond to each of two or more simultaneously presented stimuli. In the oddity procedure described above, for example, it is necessary to determine the relative frequency of responses to the odd stimulus under all possible stimulus combinations.

Similar considerations arise in studies of matching-to-sample behavior. The traditional approach to this problem is to present a new stimulus combination after each response, while reinforcing every occurrence of the appropriate response. An alternative approach is to present a new stimulus combination after each response, but to reinforce appropriate

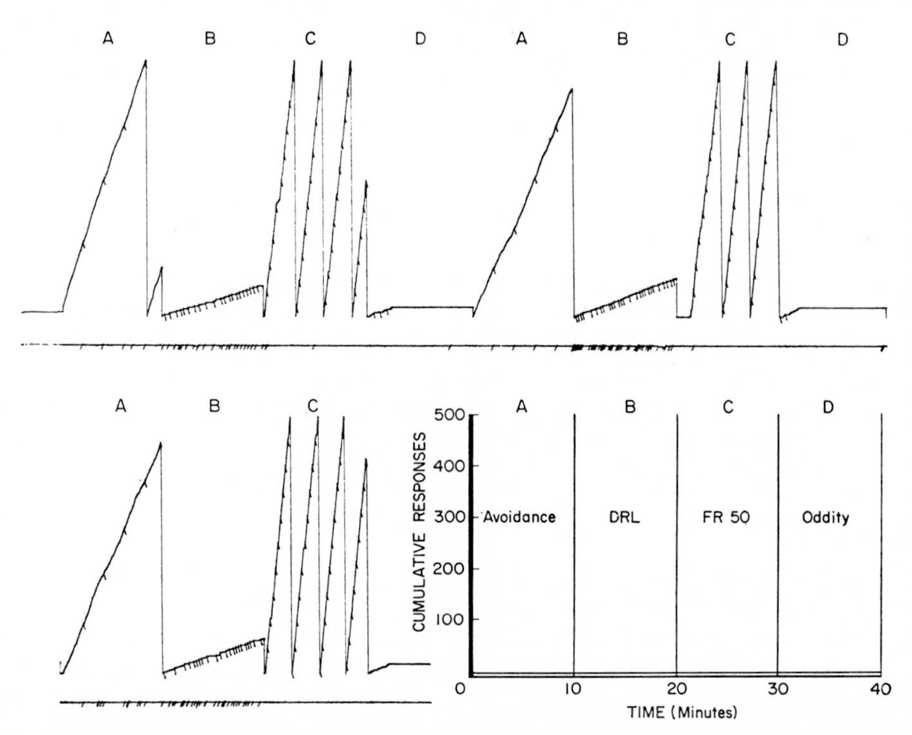

FIG. 13. Representative performance of a chimpanzee on a complex multiple schedule of reinforcement. The pen reset to the bottom of the record when 500 responses had accumulated or following each change in schedule. The recorder did not run during the time-out periods that followed each component schedule. Schedule A (Avoidance): responses on the right lever were cumulated, presentations of the blue light appear as pips on the cumulative records, and responses on the left lever are shown as pips on the lower record. Schedule B (DRL): responses on the right lever were cumulated, presentations of the stimulus light above the drinking tube are shown as pips on the cumulative records, and pressures on the drinking tube are shown as pips on the lower record. Schedule C (FR 50): responses on the center lever were cumulated, and pips on the cumulative record indicate food reinforcements. Schedule D (Oddity): responses to the odd stimulus were cumulated while responses to either of the identical stimuli are shown as pips on the cumulative record. (From Belleville *et al.*, 1963.)

responses intermittently; that is, discriminative behavior itself can be reinforced according to schedules. The studies discussed in this section show that such schedules can control not only the rate of occurrence of the discriminative behavior but also the quality of the discriminative behavior.

1. INTERMITTENT REINFORCEMENT OF ODDITY RESPONSES

The oddity procedure used by Belleville *et al.* (1963) was described above. Rohles *et al.* (1961) reported that about 95% of all responses were oddity responses when each oddity response was reinforced by food. In a subsequent experiment with one chimpanzee, oddity responses were reinforced according to a 19-response fixed-ratio (FR-19) schedule (Rohles, 1961). Every oddity response advanced the sequence of stimuli, but only the 19th oddity response was reinforced. Errors still produced a brief time out but did not advance the sequence nor count toward reinforcement. During the 42-minute session shown in Fig. 14, the chimpanzee made about 7,200 responses, including only 15 errors. Observations of this chimpanzee suggested that its performance was at least partially dependent on the sequence in which the stimuli appeared. While ap-

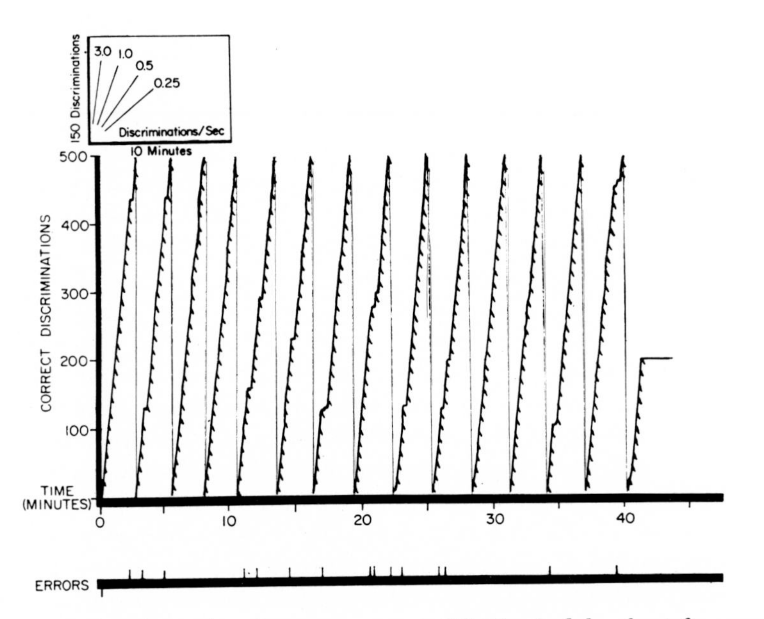

FIG. 14. Performance of a chimpanzee on an FR-19 schedule of reinforcement of oddity responses. Pips on the cumulative record indicate reinforcements. Errors are shown on the lower record. The recording pen reset to the bottom of the record when 500 responses had accumulated. (From Rohles, 1961.)

parently looking away from the display windows, the chimpanzee sometimes made as many as five consecutive oddity responses. Following an error, the animal appeared to scan the display windows when the next set of stimuli appeared. As in the experiments on concept formation by chimpanzees (Section III, A, 2), the bases for complex discriminations must be determined by experimental analysis.

2. Intermittent Reinforcement of Matching to Sample

Skinner (1950) and Ferster (1960) have described schedules of reinforcement of matching to sample in the pigeon. In Ferster's study, for example, the wall of the experimental chamber contained three translucent keys. Each key could be illuminated with red or white light. At the start of the experiment, the center key was colored, while the side keys were dark. Pecking the center key illuminated the side keys. Pecking the side key that corresponded to the color of the center key (matching response) was intermittently reinforced by food; every matching response produced a flash of light. Ferster studied matching-to-sample performance on fixed-ratio, fixed-interval, and variable-interval schedules of reinforcement. The patterns of occurrence of matching responses were similar to the patterns of occurrence of single responses on comparable schedules of reinforcement. However, the percentage of side-key responses that were matching responses was highest when matching responses were reinforced on 16- to 26-response fixed-ratio schedules. On these fixed-ratio schedules, total response rates of about 50 per minute were maintained, with 90% matching responses on the side keys.

The writer has combined two types of intermittency in a study of matching to sample in the squirrel monkey. In the experimental chamber used, three levers were mounted just below three corresponding display windows. Each window could be illuminated red or white. At the start of each sequence, the left window was illuminated, while the others were dark. Five responses on the left lever illuminated the center and right windows. Responses to the lever under the window that corresponded to the color of the left window (matching responses) produced a flash of light, and the left display window was illuminated again. Every 10th matching response was reinforced by food. Responses to the lever under the nonmatching window (errors) produced a 1-second time out. With this procedure, the monkey had to complete at least 60 responses per reinforcement (50 responses on the left lever and 10 matching responses). Figure 15 shows the performance of a squirrel monkey on this matching procedure. The monkey responded at an average rate of about two responses per second in this session and completed 600 matching responses, while making only 47 errors.

3. INTERMITTENT REINFORCEMENT OF RESPONSE SEQUENCES

In the procedures just described, the discriminative response that is reinforced depends on exteroceptive stimulus changes. Another type of procedure establishes discriminative behavior by reinforcing a response that depends on the animal's preceding responses. Examples are single-alternation and double-alternation procedures (see Chapter 5 by French). These procedures have long interested comparative psychologists because it has been assumed that efficient levels of performance depend upon the animal's symbolic processes (see Chapter 7 by Warren).

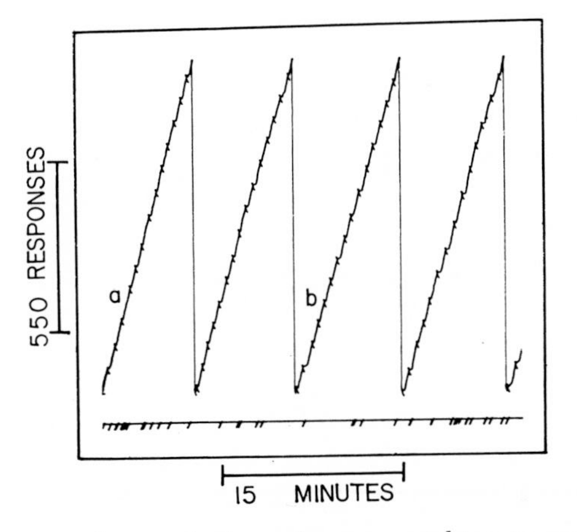

FIG. 15. Performance of a squirrel monkey on matching to sample. All responses were cumulated. Pips on the cumulative record indicate reinforcements; pips on the lower record indicate errors. The recording pen reset to the baseline when 1,100 responses had accumulated, and the recorder stopped during time-out periods. A segment in which several errors resulted in a large number of responses being required for reinforcement is shown at *a*. The completion of several ratios without an error is shown at *b*.

Ferster (1958b) studied response sequences of a chimpanzee. Pressing lever B was reinforced by food only after lever A had been pressed three times. Responses on lever B that occurred after more or fewer than three responses on lever A produced a brief time-out period. The result was that 80% of the response sequences conformed to the required sequence. The schedule was then changed so that each appropriate response sequence produced a stimulus that was occasionally followed by food. In this case, every 33rd completion of an appropriate response sequence was followed by food. With this FR procedure, 98% of the response sequences conformed to the required sequence. These results dramatically demonstrate that the level at which an animal can perform on a particular discrimination depends as much on the schedule of reinforcement as on the discriminative procedure.

These procedures for maintaining discriminative behavior with intermittent reinforcement have definite practical advantages. First, the discriminative performance can be maintained for prolonged periods of time without delivering so many reinforcers that the animal becomes satiated. Second, the rate of responding, which can be measured at the same time that discriminative performance is being measured, provides a sensitive, continuous dependent variable.

Some investigators may feel that they do not want to add the complexities of schedules of reinforcement to the complexities of discrimination procedures, but there is no real alternative. Reinforcing every response does not eliminate the schedule complexities. Investigators should always be aware that a schedule of continuous reinforcement is a schedule with its own peculiar properties, especially since it is at one extreme of the continuum of frequency and probability of reinforcement. Relatively few experiments have studied intermittent reinforcement of discriminative behavior, but results such as Ferster's (1958b, 1960) indicate that the accuracy of a discriminative performance can be improved by changing from a schedule of continuous reinforcement to a schedule of intermittent reinforcement. Precisely scheduled intermittent reinforcement may provide powerful control of each discriminative response. Indeed, the complex adaptive behavior of primates in the world outside the laboratory is usually controlled by intermittent reinforcement.

C. New Techniques for Studying Discriminative Behavior

Discussions of discriminative behavior in primates frequently raise two questions. First, what is the most complex type of discriminative behavior that can be developed in a given animal? Second, how do animals develop discriminative behavior? If the first question could be answered for a number of species, the degree of complexity attained by a given animal would provide an objective behavioral measure of its phylogenetic status. It has become increasingly apparent, however, that the answer to the first question depends critically on the answer to the second one. Without an understanding of the bases of discriminative behavior, the established upper limits of complexity are more likely to reflect the limitations of current experimental techniques than the limitations of the animal. The conclusion that a given animal cannot learn a particular discrimination may only mean that the investigator does not know how to teach it!

It has often been suggested that the acquisition of complex discriminations depends on the organism's history in acquiring simple discriminations. Skinner (1954, 1958) suggested carefully controlling this history by precisely scheduling increasingly complex discriminations. The rate at which the complexity increases should depend on the organism's performance, and it should be adjusted to minimize errors. The present section will consider two recent studies that illustrate what can be accomplished with this approach.

1. ERRORLESS DISCRIMINATION LEARNING AND TRANSFER

It has usually been assumed that the experimental extinction of errors is a necessary part of discrimination learning, but Terrace (1963a) demonstrated that pigeons could acquire a red-green discrimination without making any errors. The procedure for developing an errorless discrimination had two characteristics. First, the negative stimulus was introduced immediately after responding had been conditioned to the positive stimulus. Second, the discriminative stimuli initially differed in brightness, duration, hue, and saturation; the brightness and duration of the negative stimulus were then progressively increased until the stimuli differed only in hue and saturation. Errors occurred if the birds had much conditioning in the presence of the positive stimulus before the negative stimulus was introduced or if the stimuli differed only in hue and saturation at the start of training. Pigeons that made errors early in training continued to make occasional errors even with prolonged training; pigeons trained without errors never made errors.

In a subsequent study (Terrace, 1963b), the errorless discrimination between red and green was transferred to a discrimination between vertical and horizontal lines, also without errors. The vertical line was superimposed on the red stimulus, and the horizontal line was superimposed on the green stimulus. Then, the red and green stimuli were progressively dimmed and finally omitted. Pigeons trained in this way developed the vertical-horizontal discrimination without errors. Another group of pigeons, for which the stimuli were changed abruptly, required many sessions to develop the vertical-horizontal discrimination. When both groups were abruptly returned to the initial red-green discrimination, errors were made only by the birds that had made errors when abruptly shifted to the vertical-horizontal discrimination. Doses of chlorpromazine or imipramine that markedly disrupted the performances of pigeons trained with errors did not affect the performances of pigeons trained without errors (Terrace, 1963c).

These experiments indicate that once a bird has made errors on a discrimination it will be more likely to make more errors on that discrimination, as well as on related ones. Although these experiments were conducted with nonprimates, the results are obviously relevant to discrimination learning by other organisms.

2. Progressive Development of Matching to Sample

In one recent study with primates, Hively (1962) studied performance on a matching-to-sample procedure as a function of type of previous training on a series of similar discriminations. The subjects in this study were preschool and first-grade children. The two choice stimuli, differing in shape, size, and color, appeared in the lower section of the apparatus; the sample stimulus appeared just above them (Fig. 16). If the subject pressed the display window containing the choice stimulus that matched the sample stimulus, a new set of stimuli was presented. If the subject pressed the window containing the nonmatching stimulus, no change occurred, and the subject could respond again.

Fig. 16. Apparatus used in progressive development of matching to sample, showing a representative set of stimuli. (From Hively, 1962.)

In the progressive training procedure, the positions of the choice stimuli were not interchanged from trial to trial. At first, the sample stimulus appeared directly above the matching choice stimulus, and the nonmatching choice stimulus was not presented. Later, the nonmatching stimulus was introduced at a very low intensity, and its intensity was progressively increased until it was as bright as the matching stimulus. Finally, the sample stimulus was gradually shifted to a centered position above the choice stimuli. In the criterion procedure, the sample stimulus was centered above the choice stimuli, and the choice stimuli appeared at random in either choice window.

Training on the progressive procedure resulted in much better performance on the criterion procedure than did training only on the criterion procedure. Further, when the subjects that had been trained only on the criterion procedure were changed to the progressive training procedure, they still made many more errors than the other subjects made during their initial training on the progressive procedure.

The experiments described in this section have shown that the progressive-training approach can facilitate discrimination learning and can establish discriminative behaviors having unique characteristics. Further, they suggest that progressive discrimination training may be necessary for establishing some complex discriminations. The identification of the specific variables that determine whether or not an animal can develop a particular complex discrimination should have important implications for theories of discrimination learning and for human education.

IV. SUMMARY

The first part of this chapter emphasized new techniques for objectively analyzing aversively-controlled behavior of nonhuman primates. The behavior of primates is particularly sensitive to control by scheduled presentations of electric shock. High levels of responding can be developed and maintained by continuous-avoidance schedules or escape schedules. After continuous-avoidance training, behavior is highly resistant to experimental extinction, and stimulus generalization gradients are relatively flat. On the other hand, ongoing behavior can be totally suppressed by presenting a stimulus that regularly precedes unavoidable shock. However, when such preshock stimuli are presented to animals that have been trained on continuous avoidance, the suppressant effects may be completely reversed. Several experiments showed that, in animals with a history of avoidance training, behavior could be maintained solely by unavoidable shocks.

Psychologists have argued about the effects of aversive stimuli on behavior. Does anxiety or fear suppress behavior, or does it activate behavior? Can behavior be irreversibly suppressed by an aversive stimulus, or does such a stimulus generate behavior? The findings discussed in Section II indicate that the organism's experiences with aversive stimuli can determine whether an aversive stimulus will suppress or facilitate behavior.

The second part of this chapter emphasized the use of new scheduling techniques for studying discriminative behaviors. Different rates and patterns of responding in the presence of different stimuli were developed in primates by multiple schedules of reinforcement. Several experiments illustrated the use of simple multiple schedules to show selective effects of drugs, and the use of complex multiple schedules to assess the effects of space flight or to study concept formation. The concept-formation experiments showed that chimpanzees could respond to the common element in a series of complex stimuli; however, in one experiment the animals responded to each of the 26 specific stimuli rather than to the common element.

Discriminative behavior itself can be maintained on a schedule of reinforcement. For example, in a matching-to-sample procedure, matching responses can be intermittently reinforced. Several experiments have shown that the rate of occurrence of a discriminative response can be brought under the same type of schedule control as the rate of occurrence of a simple response. This technique enables the experimenter to maintain discriminative behavior for relatively long periods of time, and to control the rate at which discriminative behavior occurs. Further, these experiments suggested that discriminative performance could be more accurate on a schedule of intermittent reinforcement than on a continuous-reinforcement schedule; that is, the level of discrimination was a function of the schedule of reinforcement.

Other recent experiments have shown the advantages of training an organism on a carefully scheduled series of progressively more difficult discriminations. In one study, pigeons were trained without errors on a simple color discrimination, and errorless transfer to a form discrimination was also accomplished. Discriminations established without errors were more resistant to disruption than those established with errors. In another study, children were trained on a series of discriminations that progressively approached matching to sample. Without the progressive training, children made many errors on the matching procedure. Both studies indicated that once subjects had made errors, they tended to continue to make errors.

All too often the dynamic aspects of discriminative performances are neglected. It is often assumed that an organism either "knows" a discrimination or doesn't "know" it. However, dependence of accuracy of a discriminative performance on the schedule of reinforcement used to maintain it suggests that this assumption is wrong. Also, the failure of an organism to develop a particular discriminative performance is often assumed to mean that the organism cannot learn that discrimination. However, experiments described in this chapter have shown that a discriminative performance is a function of the organism's history. Further investigations of both of these variables—reinforcement schedule and training history—will be required to establish a broad empirical basis for future theories of discrimination learning.

REFERENCES

Appel, J. B. (1960). Some schedules involving aversive control. *J. exp. Anal. Behav.* **3**, 349.

Azrin, N. H., Holz, W. C., & Hake, D. (1962). Intermittent reinforcement by removal of a conditioned aversive stimulus. *Science* **136**, 781.

Azrin, N. H., Holz, W. C., Hake, D. F., & Allyon, T. (1963). Fixed-ratio escape reinforcement. *J. exp. Anal. Behav.* **6**, 449.

Belleville, R. E., Rohles, F. H., Grunzke, M. E., & Clark, F. C. (1963). Development of a complex multiple schedule in the chimpanzee. *J. exp. Anal. Behav.* **6**, 549.

Brady, J. V. (1958). Animal experimental evaluation of drug effects upon behavior. *Fed. Proc.* **17**, 1031.

Blough, D. S. (1961). Generalization and preference on a stimulus-intensity continuum. *J. exp. Anal. Behav.* **2**, 307.

Clark, F. C. (1961). Avoidance conditioning in the chimpanzee. *J. exp. Anal. Behav.* **4**, 393.

Cook, L., & Kelleher, R. T. (1962). Drug effects on the behavior of animals. *Ann. N. Y. Acad. Sci.* **96**, 315.

Estes, W. K., & Skinner, B. F. (1941). Some quantitative properties of anxiety. *J. exp. Psychol.* **29**, 390. (Reprinted in "Cumulative Record," pp. 393–404. Appleton-Century-Crofts, New York, 1959.)

Ferster, C. B. (1953). The use of the free operant in the analysis of behavior. *Psychol. Bull.* **50**, 263.

Ferster, C. B. (1958a). Control of behavior in chimpanzees and pigeons by time out from positive reinforcement. *Psychol. Monogr.* **72**, No. 8 (Whole No. 461).

Ferster, C. B. (1958b). Intermittent reinforcement of a complex response in a chimpanzee. *J. exp. Anal. Behav.* **1**, 163.

Ferster, C. B. (1960). Intermittent reinforcement of matching to sample in the pigeon. *J. exp. Anal. Behav.* **3**, 259.

Ferster, C. B., & Skinner, B. F. (1957). "Schedules of Reinforcement." Appleton-Century-Crofts, New York.

Fry, W., Kelleher, R. T., & Cook, L. (1960). A mathematical index of performance on fixed-interval schedules of reinforcement. *J. exp. Anal. Behav.* **3**, 193.

Harlow, H. F. (1951). Primate learning. *In* "Comparative Psychology" (C. P. Stone, ed.), 3rd ed., pp. 183-238. Prentice-Hall, New York.

Hearst, E. (1960). Stimulus generalization gradients for appetitive and aversive behavior. *Science* **132**, 1769.

Hearst, E. (1962). Concurrent generalization gradients for food-controlled and shock-controlled behavior. *J. exp. Anal. Behav.* **5**, 19.

Hearst, E., Beer, B. S., Sheatz, G., & Galambos, R. (1960). Some electrophysiological correlates of conditioning in the monkey. *EEG clin. Neurophysiol.* **12**, 137.

Herrnstein, R. J., & Sidman, M. (1958). Avoidance conditioning as a factor in the effects of unavoidable shocks on food-reinforced behavior. *J. comp. physiol. Psychol.* **51,** 380.

Hively, W. (1962). Programming stimuli in matching to sample. *J. exp. Anal. Behav.* **5,** 279.

Kelleher, R. T. (1957). A multiple schedule of conditioned reinforcement with chimpanzees. *Psychol. Rep.* **3,** 485.

Kelleher, R. T. (1958). Concept formation in chimpanzees. *Science* **128,** 777.

Kelleher, R. T., & Cook, L. (1959). An analysis of the behavior of rats and monkeys on concurrent fixed-ratio avoidance schedules. *J. exp. Anal. Behav.* **2,** 203.

Kelleher, R. T., & Morse, W. H. (1964). Escape behavior and punished behavior. *Fed. Proc.* **23,** 808.

Kelleher, R. T., Riddle, W. C., & Cook, L. (1963). Persistent behavior maintained by unavoidable shocks. *J. exp. Anal. Behav.* **6,** 507.

Miller, N. E. (1948). Theory and experiment relating psychoanalytic displacement to stimulus-response generalization. *J. abnorm. soc. Psychol.* **43,** 155.

Nissen, H. W. (1951). Analysis of a complex conditional reaction in chimpanzee. *J. comp. physiol. Psychol.* **44,** 9.

Rohles, F. H., Jr. (1961). The development of an instrumental skill sequence in the chimpanzee. *J. exp. Anal. Behav.* **4,** 323.

Rohles, F. H., Jr., Belleville, R. E., & Grunzke, M. E. (1961). Measurement of higher intellectual functioning in the chimpanzee. *Aerospace Med.* **32,** 121.

Rohles, F. H., Jr., Grunzke, M. E., & Reynolds, H. H. (1963). Chimpanzee performance during the ballistic and orbital Project Mercury flights. *J. comp. physiol. Psychol.* **56,** 2.

Sidman, M. (1953a). Avoidance conditioning with brief shock and no exteroceptive warning signal. *Science* **118,** 157.

Sidman, M. (1953b). Two temporal parameters of the maintenance of avoidance behavior by the white rat. *J. comp. physiol. Psychol.* **46,** 253.

Sidman, M. (1958). By-products of aversive control. *J. exp. Anal. Behav.* **1,** 265.

Sidman, M. (1960). Normal sources of pathological behavior. *Science* **132,** 61.

Sidman, M. (1961). Stimulus generalization in an avoidance situation. *J. exp. Anal. Behav.* **4,** 157.

Sidman, M., & Boren, J. J. (1957). The use of shock-contingent variations in response-shock intervals for the maintenance of avoidance behavior. *J. comp. physiol. Psychol.* **50,** 558.

Sidman, M., Herrnstein, R. J., & Conrad, D. G. (1957). Maintenance of avoidance behavior by unavoidable shocks. *J. comp. physiol. Psychol.* **50,** 553.

Sidman, M., Mason, J. W., Brady, J. V., & Thach, J., Jr. (1962). Quantitative relations between avoidance behavior and pituitary-adrenal cortical activity. *J. exp. Anal. Behav.* **5,** 353.

Skinner, B. F. (1938). "The Behavior of Organisms: An Experimental Analysis." Appleton-Century, New York.

Skinner, B. F. (1950). Are theories of learning necessary? *Psychol. Rev.* **57,** 193. (Reprinted in "Cumulative Record," pp. 39–69. Appleton-Century-Crofts, New York, 1959.)

Skinner, B. F. (1954). The science of learning and the art of teaching. *Harvard educ. Rev.* **24,** 86. (Reprinted in "Current Trends in Psychology and the Behavioral Sciences," pp. 38–58. Univer. Pittsburgh Press, Pittsburgh, Pennsylvania, 1955. Also in "Cumulative Record," pp. 145–157. Appleton-Century-Crofts, New York, 1959.)

Skinner, B. F. (1956). A case history in scientific method. *Amer. Psychologist* **11,** 221. (Reprinted in "Psychology: A Study of a Science" [S. Koch, ed.], Vol. 2, pp. 359–379. McGraw-Hill, New York, 1959. Also in "Cumulative Record," pp. 76–100. Appleton-Century-Crofts, New York, 1959.)

Skinner, B. F. (1958). Teaching machines. *Science* **128,** 969. (Reprinted in "Cumulative Record," pp. 158–177. Appleton-Century-Crofts, New York, 1959.)

Terrace, H. S. (1963a). Discrimination learning with and without "errors." *J. exp. Anal. Behav.* **6,** 1.

Terrace, H. S. (1963b). Errorless transfer of a discrimination across two continua. *J. exp. Anal. Behav.* **6,** 223.

Terrace, H. S. (1963c). Errorless discrimination learning in the pigeon: effects of chlorpromazine and imipramine. *Science* **140,** 318.

Waller, M. B., & Waller, Patricia F. (1963). The effects of unavoidable shocks on a multiple schedule having an avoidance component. *J. exp. Anal. Behav.* **6,** 29.

Weinstein, B. (1945). The evolution of intelligent behavior in rhesus monkeys. *Genet. Psychol. Monogr.* **31,** 3.

Chapter 7

Primate Learning in Comparative Perspective[1]

J. M. Warren

Department of Psychology,
The Pennsylvania State University, University Park, Pennsylvania

I. INTRODUCTION

In what ways does learning in primates differ from learning in other vertebrates? This chapter reviews the experimental data regarding learning by primate and nonprimate species with the objective of answering that question.

The review is doubly selective. It considers only those problems on which both primates and representatives of some other vertebrate taxon have been tested. Thus, there will be very little to say about maze learning since primates are seldom tested in mazes, and nothing at all to say about the Weigl oddity problem (see Chapter 5 by French) because no nonprimate has solved it.

[1]The preparation of this chapter and much of the previously unpublished research described here were supported by Grant M-04726 from the National Institute of Mental Health, U. S. Public Health Service.

II. CONDITIONING

A. Acquisition of Conditioned Responses

1. Classical Conditioning

The rate of conditioning varies markedly within the same species, even within the same organism, as a function of many experimental conditions. For example, macaque monkeys (*Macaca* spp.) learned to make anticipatory movements to a tone associated with shock after 10 paired presentations of tone followed by shock (Harris, 1943), but consistent eyelid closure was not observed after 300 combinations of light (CS) and air puff (UCS) in some of the rhesus monkeys (*Macaca mulatta*) studied by Hilgard and Marquis (1936). Indeed, the rate of conditioning of different sorts of responses varies so much within species it is impossible to show any meaningful pattern of phyletic differences.

Nor does it seem likely that other measures of conditioning will reveal coherent patterns of interspecies differences. Hilgard and Marquis compared the changes in response topography during conditioning of lid closure in dog, rhesus monkey, and man, and concluded that "the responses of dog and man are more alike than those of man and monkey" (Hilgard & Marquis, 1936, p. 198).

A similar story may be told about recent studies of the interval between the conditioned and unconditioned stimuli most favorable for rapid learning. Noble *et al.* (1959) found the optimal CS-UCS interval for fish of the genus *Mollienisia* to be 2.0 seconds. The optimal interval in man is 0.5 second (Kimble, 1961). The results obtained from *Mollienisia* were exciting since the longer optimal interval in this form suggested a behavioral disparity which might reflect the great differences in neural organization between fish and man. However, it has since been found that the optimum interval for conditioning in rhesus monkeys is also 2.0 seconds (Noble & Harding, 1963), and is even longer than 2.0 seconds for the pig (Noble & Adams, 1963).

2. Instrumental Conditioning

Goldfish, pigeons, rats, cats, dogs, squirrel monkeys (*Saimiri sciureus*), rhesus monkeys, and chimpanzees (*Pan*) have recently been tested on the Sidman (1953a, 1953b) avoidance task, in which each response postpones the delivery of shock (Appel, 1960; Behrend & Bitterman, 1963; Black & Morse, 1961; Clark, 1961; Graf & Bitterman, 1963; Kelleher & Cook, 1959; Sidman, 1955; Sidman & Boren, 1957). No interspecies differences in learning were observed. The experimenters who worked with goldfish (Behrend & Bitterman, 1963) and with chimpanzees (Clark, 1961) explicitly pointed out the absence of any important differences between learning by these species and by rats. These results are completely typical of the findings regarding avoidance learning in general. No one has seriously maintained that vertebrate species differ materially in capacity for this kind of learning.

Comparative data on the learning of simple operant responses with positive reinforcement are relatively limited. Experimenters agree that the ease or difficulty of establishing operant responses is more strongly influenced by species' structural and behavioral specializations than by their phyletic status. The details of response shaping are now seldom reported, and species can be compared only with respect to changes in the rate or accuracy of operant performance resulting from changes in the experimental situation after the operant has been stabilized.

The reasonable assumption that the details of shaping are inconsequential for comparative purposes is validated by observations such as the following: fish, chickens, ducks, pigeons, rabbits, dogs, monkeys, and chimpanzees all learn simple discriminated operants with considerable and roughly equivalent facility (Razran, 1961; Voronin, 1962). Very similar operant performances under schedules of partial reinforcement have been demonstrated in such diverse forms as the pigeon, mouse, rat, dog, cat, monkey, and chimpanzee (Ferster & Skinner, 1957).

There is little point in describing additional conditioning experiments which fail to suggest phyletic trends. The available evidence indicates that there is no systematic variation in capacity for simple classical or operant conditioning among the vertebrate species studied thus far.

B. Discrimination of Ambiguous Conditioned Stimuli

Voronin (1962) described several tests of animals' ability to develop appropriate responses when conditioned stimuli function ambiguously, as when they precede reinforcement and nonreinforcement in different contexts. These tests consistently revealed a similar pattern of interspecies differences in learning among vertebrates. For example, baboons (*Papio*) and dogs learn in 3 to 20 trials to respond to a reinforced combination of light and sound conditioned stimuli, and not to respond to unreinforced presentations of either light or sound alone. Fish failed completely to discriminate between the compound stimulus and its components presented separately. Rabbits failed to discriminate as accurately as the dogs or baboons but performed more adequately than the fish.

C. Repeated Extinctions and Reversals of Conditioned Responses

Voronin (1962) also described a series of experiments in which discriminated operants were alternately extinguished and reconditioned until the subjects learned to stop responding after a single extinction trial. The minimum number of experimental sessions required for extinction following a single failure of reinforcement was 10 for baboons, 7 for dogs, 12 for rabbits, 36 for hens, 50 for tortoises, and 68 for fish.

Comparable results were obtained in studies of serial reversals of a discriminated operant. Animals were trained to respond to one of two stimuli, and then the significance of S^D and S^{Δ} was reversed each time a criterion of learning was attained. Chimpanzees learned after two or three reversals to shift their responses appropriately in a single trial under the altered conditions. Dogs and baboons learned almost as quickly as chimpanzees. Rabbits and jackdaws showed some improvement in intertask performance, but never approached the level of one-trial learning. Tortoises and fish gave no sign of learning to learn later reversals in fewer trials than were required for the first reversal.

The findings just presented and those of many other experiments (Razran, 1961; Voronin, 1962) indicate a consistent set of differences among vertebrate species in learning certain types of tasks. These tasks all involve stimuli to which responses have been reinforced, but the subject must learn not to respond to these stimuli under specific conditions. Monkeys and chimpanzees learn much more efficiently than rabbits and nonmammalian organisms, but the performance of the primates does not surpass that of dogs. Thus, studies of conditioned-response learning have not suggested any uniquely specific characteristics of learning by primates.

III. DISCRIMINATION LEARNING

A. Single Discrimination Habits

1. Spatial Discrimination

Spatial discrimination is ordinarily a simple and unambiguous task. The subject is confronted with two alternatives, the right and left arms of a T-maze, or objects on the right and left sides of the stimulus tray in the Wisconsin General Test Apparatus (WGTA; see Chapter 1 by Meyer *et al.*), and responding to one side but not the other is consistently reinforced. Frequently the discriminanda differ only in spatial location, but rhesus monkeys (Warren, 1959a), cats (Warren, 1959b) and chickens (Warren *et al.*, 1960) learn spatial discriminations more readily if the right and left alternatives differ in visual characteristics than if they do not.

The results obtained from experimentally naive subjects of several species, trained to discriminate between stimuli differing in both position and brightness, are presented in Table I. The rhesus monkeys were trained with a correction method, and one animal in each of the two age groups failed to reach criterion within the arbitrary limit of 50 trials; the means for the monkey groups are based on the scores of the successful subjects only. In contrast, the noncorrection method was used in training the nonprimates, and each individual was tested until it reached criterion. Although a bias in favor of the monkeys results from comparing correction with noncorrection trials and from excluding the scores of the monkeys that failed to learn quickly, the ranges of individual differences within species were, in fact, so great that none of the species tested can be thought to differ reliably from any other in speed of solving the simple spatial discrimination task.

TABLE I

TRIALS REQUIRED BY EXPERIMENTALLY NAIVE SUBJECTS TO LEARN A DISCRIMINATION WITH COMBINED SPATIAL AND VISUAL CUES

Species	Age	*N*	Mean	Range	Reference
Paradise fish[a]	Adult	72	34	0–170	Warren (1960c)
Goldfish[a]	Adult	6	20	15–25	Warren, unpublished data
Chickens[a]	53 Days	20	5	0–20	Warren *et al.* (1960)
Cats[b]	Adult	38	12	0–58	Cronholm *et al.* (1960)
Horses[b]	Adult	2	7.5	7–8	Warren & Warren (1962)
Raccoon[b]	Adult	1	6	—	Warren & Warren (1962)
Rhesus monkeys[c]	15 Days	10	8	0–50+	Mason & Harlow (1958)
	45 Days	10	6	0–50+	Mason & Harlow (1958)

[a] Criterion: 18 correct in 20 noncorrection trials.
[b] Criterion: 11 correct in 12 noncorrection trials, and last 8 all correct.
[c] Criterion: 18 correct in 20 correction (rerun) trials.

2. NONSPATIAL DISCRIMINATION

In nonspatial discrimination problems, two dissimilar stimuli are each presented on either the right or left over a series of trials in a balanced irregular sequence. The subject must suppress any tendency to respond to positional cues and must consistently select one of the stimuli, for example, black rather than white, independent of its locus in space.

There is no evidence that vertebrate species differ appreciably in the rate at which experimentally naive animals learn a nonspatial discrimination. Gardner and Nissen (1948) summarized a series of experiments with cows, horses, sheep, chimpanzees, and human idiots and imbeciles tested on a simple discrimination, between a feedbox covered with a black drape and an uncovered box. The retarded humans and the chimpanzees performed at about the same level, and both types of primates learned the discrimination more slowly than the domestic animals.

Although naive primates may not differ markedly from other mammals in the rate of learning simple positional or visual discrimination habits, a recent experiment by the writer indicates that rhesus monkeys differ markedly from cats in their responsiveness to different sorts of cues in discrimination learning. Rhesus monkeys and cats were trained to the same criterion in the WGTA with the same pair of objects differing in multiple visual dimensions. The position of the objects was varied in an irregular sequence from trial to trial, and one group of each species was required to choose consistently one of the objects and to ignore spatial cues, the usual visual discrimination task. The second group, however, was required to respond to the right or left on every trial and to ignore the objects. The naive monkeys trained on the nonspatial problem learned much more quickly than those trained on the spatial discrimination (Table II). Quite the reverse was true for the cats. The cats trained

TABLE II

TRIALS REQUIRED BY EXPERIMENTALLY NAIVE RHESUS MONKEYS AND CATS TO LEARN DISCRIMINATION HABITS

	Monkeys			Cats		
Relevant cues	*N*	Mean	Range	*N*	Mean	Range
Position	5	66	19–109	8	23	6–62
Object	5	35	12–86	10	60	1–311

on the spatial task learned in less than half as many trials as those trained on the object discrimination. These findings do not imply that monkeys are inferior to cats in spatial learning; instead, they suggest that the untrained monkey responds so strongly to object cues that it is seriously distracted when these are irrelevant and only positional cues are relevant. The cat, on the other hand, has little difficulty in suppressing responses to object cues when they are opposed to spatial cues.

Harlow (1951) has pointed out that experimentally sophisticated monkeys respond more strongly to object than positional cues. The study just decribed indicates that this preferential response to object cues in the monkey exists before any formal training. Studies of maze learning by monkeys suggest that the dominance of visual over spatial cues is very general in primates. The learning curves for sparrows, rats, and monkeys trained on the Hampton Court maze are virtually identical (Bitterman, 1960, Fig. 2). Rhesus monkeys (Orbach, 1959) learn the Lashley III maze in fewer than half as many trials as rats (Lashley, 1929), but Orbach provided a clear explanation of the monkeys' superiority. "Unlike the rat in the enclosed maze, the monkeys responded during training to the visual requirements of the situation, and the learning of sequences of turns seemed completely unnecessary. At the entrance to each alley, the monkey scanned both ends, identified the next entrance, and took the appropriate path to it" (Orbach, 1959, pp. 51–52).

Species differences in initial responsiveness to visual and spatial cues in discrimination learning are not consistently correlated with problem-solving in general. Chickens, pigeons (Jones, 1954), and "maze-dull" rats (Rosenzweig *et al.*, 1960), like monkeys, all respond preferentially to visual rather than spatial cues.

Parametric comparisons of stimulus variables in discrimination learning by primates and nonprimates should reveal additional differences in the kinds of hypotheses that untrained animals of various species bring to the learning situation. Data on the relative dominance of cues for different species are needed to sensibly interpret interspecies differences in performance, but several other kinds of information are needed as well. Recent studies of the maturation of maze learning in rhesus monkeys (Zimmermann, 1963) and in kittens (Warren & Warren, unpublished) serve to illustrate this point.

The task was the closed-field intelligence test for animals, described by Rabinovitch and Rosvold (1951) and illustrated in Fig. 1. Infant rhesus monkeys and kittens were adapted to the test situation by training them to run from a fixed starting point to the goal box in the opposite corner of the square enclosure and on a series of practice problems (A through F in Fig. 1). Pretraining was continued for each individual sub-

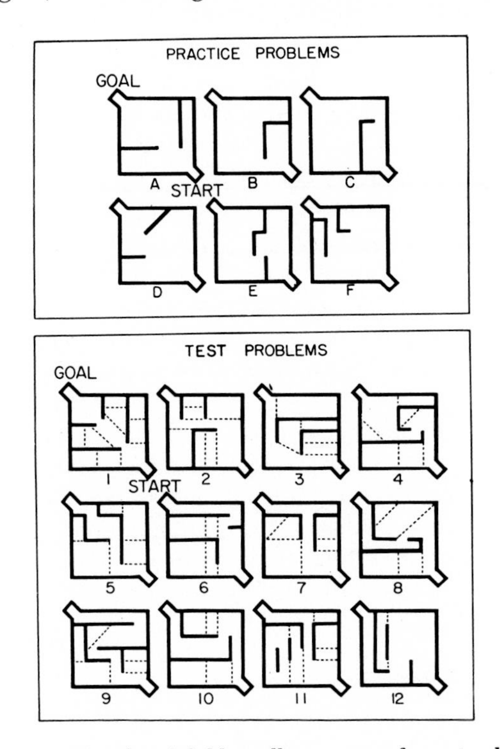

FIG. 1. The closed-field intelligence test for animals.

ject until it met a criterion of rapid and direct approach to the goal box. The animals were then tested on the standard series of 12 problems, consisting of different arrangements of barriers across the direct path from start to goal that are shown in the lower portion of Fig. 1. Errors consisted of deviations from the true path into the culs indicated by dashed lines in Fig. 1; deeper penetrations into the error zones were scored as two errors. Both the monkeys and the kittens were tested for 8 trials on each of two problems presented per day; performance was measured in terms of errors made on a total of 96 trials (twelve 8-trial problems). The rewards were meat for the kittens, and, for the baby monkeys, both milk and contact with a surrogate mother or a diaper. The results for five groups of four monkeys each, tested at 15, 45, 90, and 120 days of age, and for eight groups of cats, ranging in age from 90 to 360 days, are presented in Fig. 2. The highest mean error score for any group was that for the 120-day-old monkeys, but observations of the behavior of these animals made it clear that the objective error scores told little about their learning capacity, since they responded to the test apparatus as if it were a playground, often making a direct approach to the goal and removing the diaper without entering to end the trial. Instead, they romped about, exploiting the opportunities for play afforded by the maze, and making many more errors, without manifest interest in milk, even though they were severely deprived.

Clearly, any conclusion regarding either the ontogeny or phylogeny of learning based upon the performance of the monkeys tested at 120 days of age would be misleading, since the experimenters failed to provide a reward attractive enough to outweigh the chance to play in error zones. It is not intrinsically obvious in Fig. 2, but the data presented for cage-reared cats are also largely invalid for comparative purposes. Both alley cats (Warren *et al.*, 1961) and laboratory-reared kittens given regular opportunities to explore complex environments before being given this test make substantially lower error scores than the sample considered here.

The foregoing discussion is calculated to raise the question of the validity of interspecies differences in the measured aspects of learning performance. There are no pure, uncontaminated measures of learning proficiency; sensory, emotional, motivational, maturational, and experiential factors inevitably influence performance, and must always be considered carefully when comparing the learning observed in primates and other animal groups.

3. Probability Learning

Probability-learning experiments are characterized by the use of reinforcement ratios other than 100 : 0. Instead of reinforcing responses to one stimulus on every trial and never reinforcing responses to the other, both alternatives are reinforced in fixed proportions, such as 70 : 30 or 60 : 40, in a random sequence. Interest centers upon the sequential pattern of choices made by the subject; three types of response patterns or strategies have been described. "Matching" refers to the distribution of responses to the stimuli in a proportion approximating the ratio of reinforcement; matching may either be random and unsystematic, or reflect some clear pattern of sequential dependencies. "Maximizing" occurs when a subject consistently makes the more-frequently-reinforced choice, and never makes the less-frequently-reinforced choice. With any differ-

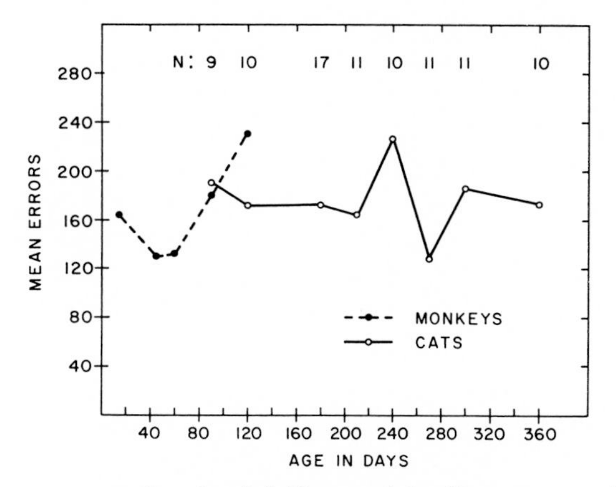

FIG. 2. Mean errors on the closed-field test of intelligence as a function of age in rhesus monkeys and cats.

ential ratio between reinforcement of the two choices, consistent performance of the more-frequently-rewarded response results in the maximum number of reinforcements possible (hence, the name for this strategy).

Fish match randomly on both spatial and nonspatial discrimination tasks (Behrend & Bitterman, 1961; Bitterman *et al.*, 1958). Pigeons maximize on spatial discrimination tasks, but fail to do so consistently on nonspatial discriminations, and match randomly under some conditions (Bullock & Bitterman, 1962; Graf *et al.*, 1964). Rats (Bitterman *et al.*, 1958) and monkeys (Meyer, 1960; Wilson, 1960; Wilson & Rollin, 1959) usually maximize. When rhesus monkeys fail to maximize, they adopt a systematic pattern of responses such as alternation after reinforcement (Wilson *et al.*, 1964); in this one instance of nonmaximizing behavior, the monkeys responded more like humans than like fish or pigeons.

The comparative results of probability-learning experiments are similar to those obtained in many conditioning studies. Monkeys behave in a clearly more adaptive manner than nonmammalian animals, but their behavior seems not to differ in any important qualitative respect from that of rats.

4. The Intermediate-Size Problem

In the intermediate-size problem, three stimuli differing in size are presented simultaneously, and the subject is obliged to choose the middle-sized rather than the largest or smallest of the three stimuli. Lashley (1938b) was unable to train rats to solve the intermediate-size problem with a reasonable amount of tuition. This task is comparatively easy for chimpanzees. Six chimpanzees learned to choose the middle-sized of three stimuli differing in the ratio of 1.6 : 1 in a mean of 135 trials (Spence, 1942). Unpublished observations by Warren indicate that the intermediate-size problem is extremely difficult for cats tested with stimuli differing in the 1.6 : 1 ratio; only one of six cats learned, and he required 1,320 trials. As the difference ratio between the stimuli was increased, more of the cats learned; but even when each stimulus was three times larger than the next smaller in the series, six cats averaged 668 trials to criterion. Under the most favorable conditions investigated so far, the cats were grossly inferior to Spence's chimpanzees in rate of learning the intermediate-size problem.

Theoretical discussions (Hull, 1952; Spence, 1942) suggest that rapid solution of problems like intermediate size depends upon precise inhibitory control of choice behavior. Progressive improvement in performance on the intermediate-size problem from rats to cats to chimpanzees lends support to recent theories that emphasize the importance of inhibitory processes in the phylogeny of learning (Harlow, 1958; Harlow & Hicks, 1957; Voronin, 1962).

The cats, like Spence's chimpanzees, were tested for transposition responses after they mastered the intermediate-size problem, to determine whether they had learned to respond to the absolute or relative size of the stimuli used in original training. If an animal had been trained initially with 100-, 160-, and 256-cm^2 squares, it was tested with 160-, 256-, and 409-cm^2 squares, and all responses were reinforced. If the subject had learned to select the middle-sized stimulus, it would select 256, but if it had learned on the basis of absolute properties of the stimuli, it would continue to select 160 even though it was the smallest of the transposition series. Table III summarizes the performances of Spence's chimpanzees and the cats on the following transposition tests: (1) 10 minutes after original learning, (2) 24 hours after original learning, (3) 24 hours after being retrained to criterion on the original problem, (4) 10 minutes after reversal learning in which the subject learned to choose the middle-sized of the original transposition set. and (5) 24 hours after reversal learning. In spite of the great difference between these species in learning, their transposition behavior is similar. Both the chimpanzees and the cats made approximately three times as many absolute as relative responses.

The obvious conclusion, that species do not vary appreciably in transfer of discrimination habits, can be generalized rather widely. A comprehensive review of studies of shape discrimination (Sutherland, 1961) reveals a remarkable degree of consistency in equivalence reactions across a wide taxonomic range. When differences in the transfer of discrimination habits are observed among species, they are frequently difficult to relate systematically to phyletic status, and appear to be more strongly influenced by differences in visual sensitivity than in learning ability.

TABLE III

PERCENTAGE OF RELATIVE AND ABSOLUTE CHOICES IN TESTS FOR TRANSPOSITION OF INTERMEDIATE-SIZE LEARNING BY 6 CHIMPANZEES AND 10 CATS

Test	Chimpanzees		Cats	
	Relative	Absolute	Relative	Absolute
1	12	88	24	74
2	23	73	14	81
3	19	74	16	83
4	39	61	33	65
5	36	64	45	53
Mean	26	72	26	71

NOTE: Rows do not add to 100% because one of the three stimuli was not intermediate in either relative or absolute terms.

Ontogenetic studies of monkeys support this conclusion. The generalization of shape-discrimination habits is essentially similar in infant and adult rhesus monkeys (see Chapter 11 by Zimmermann and Torrey in Volume II). Infant monkeys reared in a strictly-controlled visual environment generalize along a continuum on their first exposure to stimuli other than the one used in original learning of a discriminated operant (Ganz & Riesen, 1962).

To summarize the results presented in this section, experimentally naive primates evidently do not differ conspicuously from nonprimates in learning two-choice spatial or nonspatial discriminations, or in transposition or equivalence reactions. Naive primates are, however, more responsive to object than spatial cues and differ in this respect from cats, but not from chickens or pigeons. Monkeys and rats effectively maximize the number of rewards obtained in probability-learning experiments; nonmammalian species fail to do so consistently. Finally, both rats and cats are markedly inferior to chimpanzees in performance on the intermediate-size problem; this is the first task considered that appears to differentiate between the learning capacity of primates and that of other mammals.

B. Learning Sets

There are two definitions of the phenomenon of "learning how to learn" implicit in the comparative literature: (1) increased efficiency in learning repeated reversals of the same discrimination problem (see Chapter 5 by French), and (2) progressive improvement in learning consecutive nonspatial discrimination problems (see Chapter 2 by Miles). Many important questions, empirical and theoretical, about the equivalence or nonequivalence of the two definitions remain unanswered, but studies using the techniques of repeated-reversal learning and interproblem learning have yielded evidence of an orderly phyletic trend toward progressively greater learning capacity within the vertebrate series.

1. Repeated Reversals of the Same Problem

a. Spatial discrimination. Fish fail to improve in learning repeated reversals of a spatial discrimination (Bitterman *et al.*, 1958; Warren, 1960c). In fact, when 20 paradise fish were trained to criterion on each of 20 positional reversals in a T-maze, the number of trials required for learning consecutive reversals *increased* progressively (Warren, 1960c).

In contrast to the results with fish, evidence of learning to learn repeated reversals with progressively fewer errors has been obtained in experiments with turtles (Kirk & Bitterman, 1963), chickens (Warren *et al.*, 1960), and pigeons (Bitterman, 1963), as well as every mammalian species studied: rats (Dufort *et al.*, 1954), cats (Cronholm *et al.*, 1960), horses and raccoon (Warren & Warren, 1962), and rhesus monkeys (Harlow, 1949). Performance on repeated reversals of a position discrimination does not vary systematically among the mammals studied; the learning functions for rats and retarded humans are quite similar (House & Zeaman, 1959, Fig. 1).

b. Nonspatial discrimination. Turtles make progressively more errors in learning repeated reversals of a visual discrimination problem (Bitterman, 1963), their performance indicating "learning not to learn" in the same manner that fish fail on repeated reversals of a spatial habit.

But there is no evidence that the performance of primates differs substantially from that of birds or nonprimate mammals on repeated reversals of a nonspatial discrimination. Both chickens (Bacon *et al.*, 1962) and pigeons (Bitterman, 1963) show a progressive decrease in errors to criterion over repeated reversals, and eventually learn to reverse their responses very quickly. The performances of naive rats and squirrels (Rollin, 1963) and cats and rhesus monkeys (Warren, unpublished data) on a series of reversals of an object-quality discrimination in the WGTA are compared in Fig. 3. The curves for cats and monkeys are nearly identical. The performance of the rats and squirrels is inferior to that of the cats and monkeys, but chimpanzees (Nissen *et al.*, 1938; Schusterman, 1962) often do so badly on the first few reversals of a visual discrimination that it would be impossible to substantiate a claim that primates are more proficient on this kind of task than other mammals or birds.

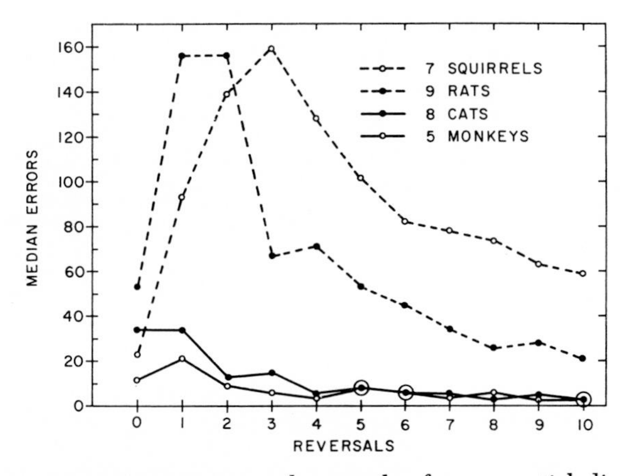

FIG. 3. Errors to criterion on repeated reversals of a nonspatial discrimination by squirrels, rats, cats, and rhesus monkeys.

2. INTERPROBLEM LEARNING

Several investigators have demonstrated the formation of learning sets by rats (Koronakos & Arnold, 1957; Weaver & Michels, 1961; Wright *et al.*, 1963), but underestimated the rat's capacity for interproblem learning because they used 2-dimensional pattern-stimuli instead of 3-dimensional objects. It is well known that rhesus monkeys form learning sets much more slowly when tested with patterns than with objects (see Chapter 1 by Meyer *et al.*). More satisfactory data for comparative purposes have been obtained by Tyrrell (1963), who trained rats in a modified WGTA to displace 3-dimensional objects with their heads. The objects were like those typically used in learning-set experiments with carnivores and primates. The performance of Tyrrell's rats on 80 50-trial discrimination problems was similar to that of cats tested under similar conditions by Warren and Baron (1956). Both species showed considerable intraproblem learning within each 20-problem block, and the performance of both species improved markedly from block to block. Initial stimulus preferences strongly affected the performance of both the rats and the cats, resulting in poorer performance on problems in which the subject made an error on trial 1 than on problems in which the subject made a correct response on this trial. This effect was greater for the cats; so if there is a difference between species under these conditions of testing, it is in favor of the rat. There is no convincing evidence that raccoons (Johnson & Michels, 1958; Shell & Riopelle, 1957) form learning sets more efficiently than cats or rats. The performance of these non-primate mammals is only slightly superior to that of pigeons tested under similar conditions (Zeigler, 1961).

By far the most conspicuous variation in capability for forming learning sets occurs among the primates. Interproblem learning over 6-trial discrimination problems by four rats and three squirrels (Rollin, unpublished), cats (Meyers *et al.*, 1962), marmosets (*Callithrix*) and rhesus monkeys (Miles & Meyer, 1956), and squirrel monkeys (Miles, 1957) is plotted in Fig. 4. All of the species studied show an improvement in

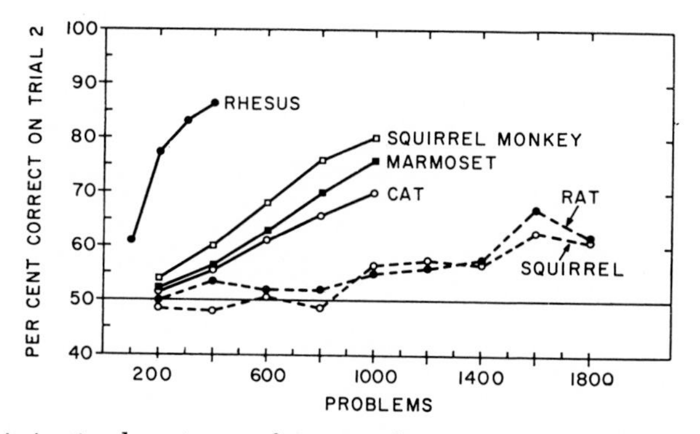

FIG. 4. Discrimination-learning-set formation by primates, carnivores, and rodents.

interproblem learning, but at markedly different rates, and attain quite different asymptotic levels of performance. The rodents' curves remain at essentially chance levels over the first 800 problems, and indicate only very slow and gradual improvement over the next 1,000 problems, the rats and squirrels averaging 64.5 and 61.3% correct responses on the last 200 problems. Their terminal level of performance is lower than that attained by cats or monkeys in many fewer problems. The rhesus monkeys are obviously superior to the squirrel monkeys, marmosets, and cats. Learning-set curves for cebus monkeys (*Cebus albifrons*) and spider monkeys (*Ateles geoffroyi*) (Shell & Riopelle, 1958) would fall between those for squirrel monkeys and rhesus monkeys in Fig. 4.

In general, the available data suggest that the phylogenetic development of capacity for learning-set formation in mammals is best described as a continuous ʃ-function, with no sharp discontinuities between adjacent taxa, but with marked quantitative differences between the extremes of the distribution. At the lower end of the scale, rats perform as well as cats or raccoons under favorable conditions (50-trial problems), but not under more rigorous circumstances of testing (6-trial problems). The transition from the carnivore to the primitive-primate level of performance is not abrupt. Taking into consideration the many data (e.g., Warren *et al.*, 1963) showing a relation between the number of differential cues available to an animal and the rate of learning-set formation, the slight superiority of the marmosets over the cats seen in Fig. 4 can be attributed very largely to the presence of color vision in marmosets (Miles, 1958) and its absence (Meyer *et al.*, 1954) or its, at best, extremely poor development (Clayton, 1963) in cats.

The steeply-rising portion of our metaphorical ogive corresponds to the rapid and steady improvement in learning-set acquisition through the New World monkeys, to the high level represented by the rhesus monkey (Harlow, 1959) and other Old World forms like the sooty mangabey (*Cercocebus torquatus*) (Behar, 1962). The upper horizontal limb of the ʃ represents the absence of any major difference in interproblem learning between macaques and anthropoid apes (Fischer, 1962; Harlow, 1959; Hayes *et al.*, 1953; Rumbaugh & Rice, 1962).

How long the currently tidy phylogenetic sequence in learning-set proficiency will endure is a matter for conjecture; Crawford (1962) has recently reported that cynomolgus monkeys (*Macaca irus*) were inferior to spider monkeys in discrimination-learning-set formation and in learning to reverse visual discriminations in response to changed locations of the stimuli.

Primate learning, as measured by the formation of learning sets, differs quantitatively but not qualitatively from learning by rodents and carnivores. Further differences between primates and other mammals have been demonstrated in respect to forming sets for discrimination reversal to a sign, and transfer of the effects of repeated-reversal training.

Rhesus monkeys learn to reverse discriminations without error when reversal of the stimulus-reward relation is signaled by a change in the color of the tray on which the discriminanda are presented (Riopelle & Copelan, 1954). Experiments comparing the effectiveness of food and nonfood signs in reversal learning were performed with cats (Warren, 1960a) and with rhesus monkeys (Warren, 1960e). Apparatus, stimulus objects, and trial procedures were identical for both species. The stimulus objects were presented on a white tray during the discrimination phase of each problem, and the white tray was retained during the reversal phase for the no-cue groups of monkeys and cats. For the cue groups, however, reversal was signaled by replacing the white tray with a black one before the first reversal trial. All of the subjects were trained to criterion on 80 discrimination and reversal problems.

The final levels of proficiency are shown in Fig. 5, which presents intraproblem learning curves for the last block of 20 problems. The curves for the monkeys are the typical discontinuous functions expected from experimentally sophisticated monkeys. The no-cue group averaged 2% and 88% correct responses on the first and second reversal trials. The cue group responded to the change in the color of the test tray as a re-

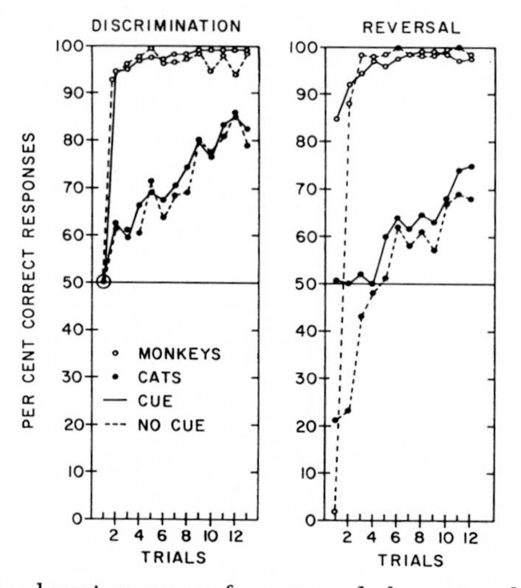

FIG. 5. Intraproblem learning curves for cats and rhesus monkeys trained on discrimination reversal to food and nonfood signs.

versal cue and made 85% correct responses on the first reversal trial. In other words, changing the color of the test tray eventually becomes as effective a cue for reversal as nonreward on the first reversal trial for monkeys.

In both discrimination and reversal learning, the curves for the cats indicate a much more gradual and continuous improvement than is seen in the monkeys. It would be misleading to compare the functions for the cats and monkeys in absolute terms because the previous histories of the groups were not identical, and the cats, of course, could not get color cues from the stimulus-objects as the monkeys could. But the cue cats had precisely the same number of experiences with the exchange of a black for a white tray as a cue for reversal, and the fact that they were much more successful than the no-cue cats on reversal trials 1 and 2 shows that they were sensitive to the tray cue. Therefore, the gross inferiority of the cue cats to the cue monkeys indicates a substantially superior ability of monkeys to form a set for discrimination reversal to a sign. Indeed, the chance performance of the cue cats over the first four reversal trials, and their failure to differ reliably from the no-cue cats in number of errors to criterion, suggest that changing the color of the test tray has the effect of a distracting stimulus rather than of a cue (Warren, 1960a).

Chimpanzees trained on repeated reversals of three visual discrimination habits showed very great facilitation when subsequently tested on a series of 6-trial discrimination problems, averaging about 90% correct on the first 30 problems (Schusterman, 1962). The transfer effect was specific to the training on reversals, since a group trained to alternate between objects performed no better than naive animals on the 6-trial problems. Schusterman attributed the superiority of the reversal group's learning-set formation to these animals' acquisition of a "win-stay, lose-shift" strategy with respect to objects during reversal training (see Chapter 3 by Levine).

I have performed a very similar experiment with three groups of cats. The control group of 11 cats, without any prior training, was trained to a criterion of 10 consecutive correct responses on each of 80 discriminations between stimuli differing in many visual dimensions. Two experimental groups of cats were trained on the same discrimination problems as the controls, after they had been pretrained on tasks calculated to produce positive or negative transfer to the learning-set series. Eight cats were pretrained on 60 reversals of a visual discrimination habit with a single pair of objects. As in learning-set training, only responses to objects were consistently rewarded; so this group, the object group, was expected to show positive transfer. The other experimental group, the position group, consisted of eight cats pretrained on 60 reversals with the same objects used in pretraining the object group. The positions of the objects were varied randomly from right to left as in a visual discrimination, but object cues were irrelevant, and responding consistently to the right or left was rewarded. Thus, the position group was trained to ignore visual cues in favor of positional cues, and should have formed habits that would interfere with subsequent learning-set formation.

Both groups learned to solve the repeated reversals with increasing efficiency and averaged about 4 errors to a criterion of 10 consecutive correct responses on the last 10 reversals presented. A portion of the interreversal learning curve for the object group is shown in Fig. 5. The median number of trials in pretraining was 1,202 for the position group and 1,458 for the object group.

When the two experimental groups were tested on the series of 80 nonspatial discrimination problems, however, their performances did not differ reliably from one another nor from that of the control group; all three groups showed a progressive reduction in the number of trials to criterion, but there was no evidence of differential transfer effects. The implication of these results is clear. Extended training on repeated reversals failed to influence subsequent learning-set performance of cats, although similar pretraining resulted in conspicuous positive transfer in Schusterman's chimpanzees. This discrepancy suggests that primates and nonprimates may differ in the types of strategies they develop on objectively similar discrimination tasks. Further research on the transfer of learning sets is badly needed.

The major points made in the discussion of learning-set formation by primates and other vertebrates may be summarized quite succinctly. Mammals and birds differ from fish and reptiles in being able to learn repeated discrimination reversals in progressively fewer trials, but primates are not markedly more proficient than other mammals or birds. The variation among mammals in capacity for interproblem learning, the capacity to learn consecutive nonspatial discrimination problems with increasing efficiency, appears to be quantitative rather than qualitative. All of the species tested have shown some capacity for learning to learn, but no nonprimate mammal yet tested has approached the level of one-trial learning observed in the rhesus monkey. Cats fail to learn discrimination reversal to a sign within a number of problems sufficient for monkeys to form such a set, and cats fail to transfer training on repeated reversals of a discrimination habit to an interproblem-learning task. Chimpanzees trained on repeated reversals of three problems show very strong positive-transfer effects.

IV. DOUBLE ALTERNATION

The subject must learn to make a series of right and left responses in a constant temporal sequence, typically RRLL, in order to solve the double-alternation problem (see Chapter 5 by French). Successful double alternation was once presumed to occur in the absence of differential exteroceptive or kinesthetic cues. In other words, it was believed to depend upon symbolic or representational processes (Hunter, 1928) analogous to counting in verbal humans (Gellermann, 1931b), and "higher" than those involved in the solution of single alternation (RLRL). Single alternation was assumed to be a simple problem that could be learned readily on the basis of differential kinesthetic cues. The status of double alternation as a measure of symbolic behavior was supported by the phyletic evidence. Rats failed to learn double alternation in the temporal maze (Hunter, 1920) or learned slowly after elaborate preliminary training (Hunter & Nagge, 1931). Raccoons (Hunter, 1928), cats (Karn, 1938; Karn & Patton, 1939), and dogs (Karn & Malamud,

1939) performed more adequately than rats, but were inferior to rhesus monkeys (Gellermann, 1931a, 1931c).

This elegant hierarchical order of capacity in mammals rested on a very flimsy foundation of facts. Most of the species studied were represented by only one or two subjects, and it was impossible to estimate the range of individual differences within species in capacity for double-alternation learning. More adequate samples of rhesus monkeys (Warren & Sinha, 1959), cats (Warren, 1961), and raccoons (Johnson, 1961) have recently been tested on double alternation in the WGTA under very similar experimental conditions. All were trained on sequences of four responses, with responses in RRLL sequence rewarded as described in detail by Stewart and Warren (1957). The results are shown in Fig. 6 as

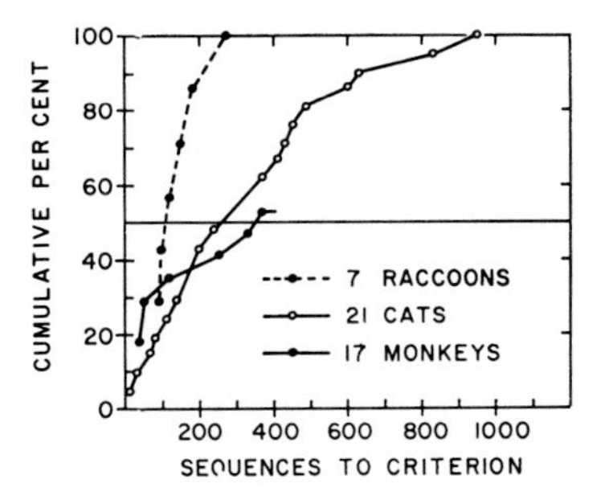

FIG. 6. Percentages of raccoons, of cats, and of rhesus monkeys having met a criterion of double-alternation learning.

cumulative percentages of subjects meeting criterion (80% correct over 50 sequences) after the number of sequences indicated on the abscissa. The curve for the monkeys ends abruptly because the animals were tested on a maximum of 450 sequences. The median number of sequences to criterion was 110 for the raccoons, 270 for the cats, and 370 for the monkeys. Individual differences within species, however, were great enough to preclude any inference of reliable differences between species. Some monkeys and cats solve double alternation almost immediately, while others require substantially more extensive training.

The distribution of double-alternation scores for the rhesus monkeys suggests a resolution of an annoying discrepancy in the primate literature. No one, to the best of my knowledge, has ever succeeded in replicating Gellermann's (1931c) results on rapid double-alternation learning by monkeys. Extremely slow and imperfect learning has been observed in small samples of rhesus (Leary *et al.*, 1952; Warren *et al.*, 1957) and cynomolgus (Warren, unpublished data) macaques, and gibbons (*Hylobates*) (Schusterman & Bernstein, 1962). The pattern of variation among the rhesus monkeys studied by Warren and Sinha suggests that Gellermann was unusually fortunate in his sampling and worked with representatives of the rather small minority of monkeys which are highly competent at double-alternation learning, and that later workers failed to replicate his results because their samples included only monkeys of average competence and none of the gifted minority.

The data presented in Fig. 6 demonstrate that primates and carnivores do not differ significantly in learning of double-alternation sequences of four responses, and Stewart and Warren (1957) have shown that some cats, like Gellermann's monkeys, can perform adequately on more extensive double-alternation sequences of eight responses. The observations that rabbits can learn double alternation in the WGTA at least as rapidly as cats (Livesey, 1964) and that rats can learn double alternation readily in the Skinner box (Schlosberg & Katz, 1943; G. Heise, personal communication) complete the demolition of the alleged correlation between phyletic status and double-alternation capability in mammals.

There are additional grounds for rejecting claims that double alternation is a test of symbolic behavior. Rhesus monkeys deprived of most of their cerebral cortex are not significantly impaired on this task relative to intact controls (Warren *et al.,* 1957). Yamaguchi and Warren (1961) found that cats learn double alternation reliably more rapidly than single alternation, and presented detailed analyses indicating that double alternation is learned as a highly specific serial order (see Chapter 5 by French). Therefore, the relative difficulty of double and single alternation varies among species as a function of characteristic differences in sequential-response preferences, and not as a function of general capacity for problem solving.

It is by now patent that the only reason for treating double alternation in a separate section of the chapter is historical precedent. Empirically, the results obtained in testing primates and other mammals on this task would dictate its inclusion in Section III, A, since it is no more successful than simple discrimination problems in differentiating the performance of primates from that of other mammals.

V. DELAYED RESPONSE

Like double alternation, delayed-response performance was once thought to imply the presence of symbolic or ideational processes in animals. This belief was also supported by data showing that delayed-response performance was correlated with phyletic status among the mammals (see Chapter 4 by Fletcher). Both notions were severely criticized by Maier and Schneirla (1935), who pointed out that every species studied could delay successfully for hours, indeed for days, if tested with distinctively different response-alternatives, and that no mammal could long delay its responses if the alternatives lacked distinctiveness.

Until recently, conservatives could object to the heterogeneous tests that Maier and Schneirla accepted as valid measures of delayed-response capability, and maintain that, if they were tested under strictly comparable conditions, nonprimate mammals would be found inferior to primates in terms of the maximum delay possible in delayed-response tests. But raccoons (Michels & Brown, 1959) and cats (Warren *et al.*, 1962) can successfully delay their responses for at least 40 seconds in the WGTA under conditions approximating those of experiments with primates. Although no effort was made to determine precisely the upper limit of the carnivores' capacity for delaying responses, the high levels of performance observed with 40-second delays suggest that primates and carnivores may not differ materially in respect to the maximum delay possible in the WGTA.

Primates and nonprimates also seem not to differ in some other aspects of delayed-response proficiency. Intersubject variability among 17 rhesus monkeys trained on 5-second delayed response was so great that French and Harlow (1962) published individual learning curves for their animals, because they felt that any single learning function based upon pooled data would be misleading. The distribution of individual differences was similar to the distributions of individual scores of monkeys and cats on double alternation (see Section IV). Comparison of the individual monkeys' delayed-response data provided by French and Harlow with the data for 10 cats trained on the same problem (Warren, 1964a) reveals that 6 of 17 monkeys failed to reach 80% correct responses in as few trials as the median cat. Primates and carnivores apparently do not differ substantially in capacity for rapid learning of the delayed-response task, a conclusion supported by the observation that under particularly favorable conditions both cats and dogs learn delayed response in a single testing session (Lawicka, 1959).

It has also proved impossible to differentiate between primates and other mammals in terms of the subjects' behavior during the delay period. New World monkeys adopt overt bodily orientations during the delay interval (French, 1959; Meyers *et al.*, 1962; Miles, 1964); cats and dogs do not (Lawicka, 1959; Meyers *et al.*, 1962). Thus, there is no consistent difference between primates and carnivores in (1) reliance on bodily orientation during the delay period, (2) rate of learning the delayed-response task, or (3) length of maximum delay. But level of performance at a given delay seems to differentiate at least some species tested under at least one standard procedure (see Chapter 4 by Fletcher).

Continuing behavioral analysis has removed delayed response progressively further from the category of "symbolic" behavior. Harlow (1951) interpreted delayed response as a type of memory test, and stated, "Successful delayed response performance is possible only if an animal learns discrimination *reversal* problems in a *single* trial to an implicit (secondarily reinforced) reward and retains this learning for a period defined by the delay" (Harlow, 1951, p. 229). French (1959) showed that this hypothesis attributes too much to an animal that solves delayed response, since he demonstrated that squirrel monkeys learn only to adopt bodily orientations toward the baited locus in delayed response, unless forced to do otherwise by special experimental procedures. Harlow's explanation of delayed response in terms of one-trial reversal learning fails completely to apply to delayed-response learning by cats and raccoons, because there is no evidence of consistent one-trial *discrimination* or *reversal* learning in these forms (see Section III, B). Konorski and Lawicka (1964) have shown that many features of delayed-response performance in carnivores may be explained satisfactorily by treating delayed response as an instance of trace conditioning. (Also see Chapter 4 by Fletcher for a new analysis of the delayed-response problem.)

VI. MULTIPLE-SIGN LEARNING

The capacity for solving complex multiple-sign problems (see Chapter 5 by French) is not restricted to primates. Pigeons have solved the conditional-discrimination and matching-to-sample problems (Blough, 1956, 1957, 1959; Ferster, 1960; Ferster & Appel, 1961; Ginsburg, 1957; Skinner, 1950; Zimmerman & Ferster, 1963), and canaries have solved the oddity problem (Pastore, 1954). Rats have solved both oddity and conditional-discrimination tasks (Lashley, 1938a; North *et al.*, 1958; Wodinsky & Bitterman, 1953). Manifold differences in experimental procedure, however, preclude direct comparisons between the experiments with birds and rats and those with primates.

Cats have been trained on multiple-sign problems in the WGTA under conditions very much like those of experiments with rhesus monkeys; the remainder of this section is concerned with the comparison of oddity and conditional-discrimination learning by rhesus monkeys and cats, tested under very similar conditions.

A. Oddity

Many cats fail to master the oddity problem (Boyd & Warren, 1957; Warren, 1960b), even when given many more trials than are required for learning by rhesus monkeys (Meyer & Harlow, 1949). Of the cats that solve a single oddity problem, most fail to generalize the oddity principle. On the other hand, one cat formed a highly efficient set for the solution of oddity problems, and attained a final level of performance within the range of performance of rhesus monkeys trained under similar conditions (Warren, 1960b).

The proportion of monkeys that learn the generalized oddity principle is far higher than the proportion of cats, but some individuals of both species can reach the same final level of efficiency.

B. Conditional Discrimination

In order to determine whether nonprimate mammals could discriminate stimuli presenting multiple, ambivalent cues, Joshi and Warren (1959) tested cats on a complex conditional-discrimination problem under conditions almost identical to those of an earlier experiment with rhesus macaques (Noer & Harlow, 1946). The stimuli were large white (W), large black (B), small white (w), and small black (b) wooden objects, which were paired in four configurations (+ and — denote the positive and negative stimulus in each pair):

1. b+ and w—
2. W+ and B—
3. w+ and W—
4. B+ and b—

The final stage of training required the subjects to respond appropriately to all four configurations, presented in random sequence within a single testing session. Each object was positive in one context and negative in another, so that accurate performance depended on varying responses to each object in terms of the other discriminandum presented on any particular trial. The final stage was reached through a sequence of part problems which is shown in the second column of Table IV. This sequence and most other features were identical to those in the experiment with monkeys. The criterion of learning at each step in training was 45 correct responses in 50 consecutive noncorrection trials (2 days of testing).

Table IV also gives the median number of errors to criterion made by the monkeys and cats at each stage of learning. Medians must be used because two of the monkeys required special remedial training to learn Test VII and three monkeys failed to reach criterion on Test VIII. All of the cats learned every test, and none required any special training. The monkeys made substantially fewer errors than the cats in learning the first six tests, but the monkeys made more than 10 times as many errors as the cats on Test VIII, every cat making fewer errors than any of the monkeys at this stage. Joshi and Warren speculated that these results were artifactual in the sense that they were specific to the conditions imposed by the particular experimental design employed and possibly reflected species differences in the way that cats and monkeys learn this one particular problem. No claim that cats are peculiarly gifted in capacity for solving complex conditional-discrimination problems was made.

As a check on the general validity of the paradoxical findings just given, groups of rhesus monkeys (Warren, 1960d) and cats (Warren, 1961) with extensive experience in visual-discrimination learning were trained on a conditional-discrimination problem, without the sequence of pretraining used in the above experiment. The stimuli were a square

TABLE IV

DISCRIMINATION OF AMBIVALENT CUES BY RHESUS MONKEYS AND CATS

Test	Configurations	Median errors	
		6 Monkeys	6 Cats
I	1	6	55
II	2	17	112
III[a]	1, 2	8	43
III[b]	1, 2	2	7
IV	3	8	35
V	4	13	57
VI[a]	3, 4	26	211
VI[b]	3, 4	3	18
VII	2, 3, 4	44	30
VIII	1, 2, 3, 4	428	40

[a] Trials presented in alternate groups of five.
[b] Trials presented in balanced irregular sequence.

and a triangle. When these forms were both large (100 cm^2 in area), one was correct; but when they were both small (12.5 cm^2), the other was correct. From the outset of training, the size of the pair of figures was varied from trial to trial in an irregular sequence. Each size was presented on half the trials in a given test session, and the positive stimulus appeared equally often on the right and on the left. Both the monkeys and the cats were trained to a criterion of 20 correct responses within a block of 24 trials.

Figure 7 shows the rates of learning. Seven of the 13 monkeys reached criterion in the first 24 trials, and none required more than 48 trials to learn. Median trials to criterion for the cats was 180; however, four cats' scores fall within the range for the monkeys. These data reveal a marked quantitative superiority of rhesus monkeys over cats in learning a single conditional discrimination, but the overlapping ranges of individual scores demonstrate that the difference between these species is not all-or-none.

Nonprimates have not yet shown interproblem learning, however. Rats (Lashley, 1938a) and cats (Warren, 1964a) have shown no interproblem improvement in performance over a series of conditional-discrimination problems. Rhesus monkeys, in contrast, form conditional-discrimination-learning sets with considerable facility (Warren *et al.*, 1963). The performance of four rhesus monkeys tested on 60 24-trial conditional discriminations between geometrical forms, with a difference in surface area (36 cm² versus 81 cm²) as the conditional cue, is sum-

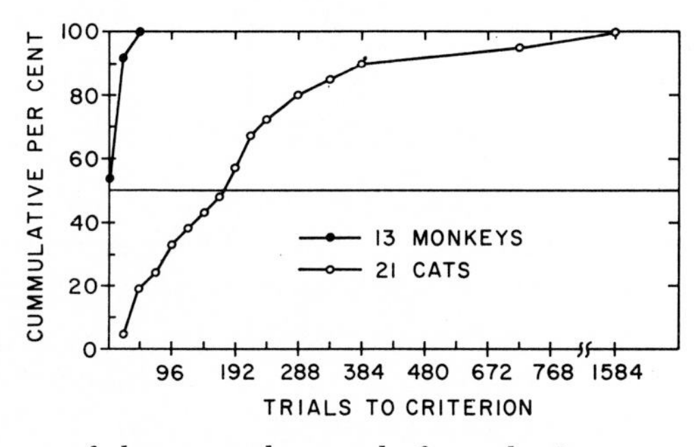

FIG. 7. Percentages of rhesus monkeys and of cats having met criterion on a conditional-discrimination problem.

marized in Fig. 8. Significant interproblem learning was observed; the monkeys made 66% and 80% correct responses on the first and last 10 conditional discriminations. The intraproblem learning curves show that the animals learned the conditional problems slowly and gradually in comparison to the rapid, discontinuous solution of simultaneous discrimination problems by monkeys.

But further research (Warren, 1964b) indicates that Fig. 8 provides only a minimal estimate of the sophisticated rhesus macaque's capacity

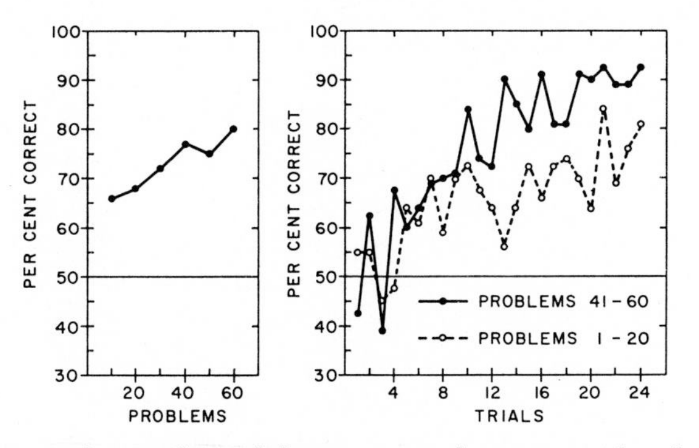

FIG. 8. Formation of a conditional-discrimination-learning set by rhesus monkeys. Left: Interproblem learning. Right: Intraproblem learning.

for solving conditional-discrimination problems. Much more rapid learning is observed when, for example, differences in color or orientation of the stimuli in the horizontal and vertical planes are made the conditional cues. The effect is probably related to the greater discriminability of these as opposed to size cues (see Chapter 1 by Meyer *et al.*).

The available evidence points toward the same conclusion about conditional-discrimination learning as about simple discrimination learning: nonprimate mammals may learn single conditional-discrimination problems at least as readily as primates, but primates clearly surpass all other forms in their capacity for rapid interproblem learning.

VII. CONCLUSIONS

No one can yet provide a definitive answer to the question "In what ways does learning in primates differ from learning in other vertebrates?" Far too little is known about learning in nonprimate species and most species of primates, particularly the more primitive primates. Also, it is safe to assume that the most favorable conditions for learning, and hence the limits of learning and problem-solving ability, remain to be established in even such intensively-studied species as the rat, rhesus monkey, and chimpanzee.

One can list with confidence the ways in which learning in primates does *not* differ from learning in other taxa: 1. The evidence fails to indicate any meaningful variation among vertebrates in the acquisition of single simple conditioned responses or discrimination habits by experimentally naive subjects. 2. Mammals, both primates and nonprimates, differ from nonmammalian forms in probability learning and in reversal learning. 3. Primates and nonprimate mammals do not differ qualitatively in their capacity for forming discrimination-learning sets, for performing delayed response or double alternation, or for solving individual conditional-discrimination or oddity problems.

This leaves a surprisingly short list of learning problems on which primates appear to be superior to nonprimate mammals: 1. Old World primates are *quantitatively* superior to nonprimate mammals in the speed with which they form learning sets, in the asymptotic level of performance attained on learning-set problems (no nonprimate animal has yet shown one-trial solution of visual discrimination problems), and in the rate at which they solve exceptionally difficult discrimination problems such as intermediate size. 2. Chimpanzees are qualitatively superior to cats in the generalization and transfer of learning sets (see Section III, B, 2). 3. Rhesus monkeys are qualitatively superior to cats in learning discrimination reversal to a sign and in forming sets for conditional-discrimination learning.

These conclusions must be regarded as tentative for two reasons. Generalizations of negative results are very hazardous; it is quite conceivable that dogs or dolphins would succeed where cats have failed, or that cats tested under more appropriate conditions would perform more adequately than the groups discussed in this chapter. In either event, the suggested qualitative differences between primates and other mammals would vanish. In the second place, the very similar learning-set performance of cats and marmosets raises considerable doubt about differences between orders in learning capacity. Differences between New World monkeys and nonprimate mammals probably will be less impressive on complex learning-set problems than are differences between cats and rhesus macaques or chimpanzees. Learning performances of primitive primates, like the lemur, probably resemble those of carnivores more than those of advanced primates.

The data summarized in this chapter point toward an important conclusion about learning theory and comparative psychology. The qualitative differences in learning between mammals and nonmammals are much greater and more significant than learning theorists (Hull, 1945; Skinner, 1956) have recognized. On the other hand, the substantial qualitative similarities in learning by mammals in general, as opposed to nonmammals, clearly indicate that the differences in learning between primates and other mammals are not so great as they seemed to many primatologists a few years ago.

REFERENCES

Appel, J. B. (1960). Some schedules involving aversive control. *J. exp. Anal. Behav.* **3**, 349.

Bacon, H. R., Warren, J. M., & Schein, M. W. (1962). Non-spatial reversal learning in chickens. *Anim. Behav.* **10**, 239.

Behar, I. (1962). Evaluation of cues in learning set formation in mangabeys. *Psychol. Rep.* **11**, 479.

Behrend, Erika R., & Bitterman, M. E. (1961). Probability-matching in the fish. *Amer. J. Psychol.* **74**, 542.

Behrend, Erika R., & Bitterman, M. E. (1963). Sidman avoidance in the fish. *J. exp. Anal. Behav.* **6**, 47.

Bitterman, M. E. (1960). Toward a comparative psychology of learning. *Amer. Psychologist* **15**, 704.

Bitterman, M. E. (1963). Species-differences in learning: fish, reptile, bird, and rat. Paper read at Amer. Psychol. Assn., Philadelphia, Pennsylvania.

Bitterman, M. E., Wodinsky, J., & Candland, D. K. (1958). Some comparative psychology. *Amer. J. Psychol.* **71**, 94.

Black, A. H., & Morse, Patricia (1961). Avoidance learning in dogs without a warning stimulus. *J. exp. Anal. Behav.* **4**, 17.

Blough, D. S. (1956). Technique for studying the effects of drugs on discrimination in the pigeon. *Ann. N. Y. Acad. Sci.* **65**, 334.

Blough, D. S. (1957). Some effects of drugs on visual discrimination in the pigeon. *Ann. N. Y. Acad. Sci.* **66,** 733.

Blough, D. S. (1959). Delayed matching in the pigeon. *J. exp. Anal. Behav.* **2,** 151.

Boyd, B. O., & Warren, J. M. (1957). Solution of oddity problems by cats. *J. comp. physiol. Psychol.* **50,** 258.

Bullock, D. H., & Bitterman, M. E. (1962). Probability-matching in the pigeon. *Amer. J. Psychol.* **75,** 634.

Clark, F. C. (1961). Avoidance conditioning in the chimpanzee. *J. exp. Anal. Behav.* **4,** 393.

Clayton, K. N. (1963). Successful performance by cats on several color discrimination problems. *Amer. Psychologist* **18,** 407. (Abstract.)

Crawford, F. T. (1962). Reversal learning to spatial cues by monkeys. *J. comp. physiol. Psychol.* **55,** 869.

Cronholm, J. N., Warren, J. M., & Hara, K. (1960). Distribution of training and reversal learning by cats. *J. genet. Psychol.* **96,** 105.

Dufort, R. H., Guttman, N., & Kimble, G. A. (1954). One-trial discrimination reversal in the white rat. *J. comp. physiol. Psychol.* **47,** 248.

Ferster, C. B. (1960). Intermittent reinforcement of matching to sample in the pigeon. *J. exp. Anal. Behav.* **3,** 259.

Ferster, C. B., & Appel, J. B. (1961). Punishment of SΔ responding in matching to sample by time out from positive reinforcement. *J. exp. Anal. Behav.* **4,** 45.

Ferster, C. B., & Skinner, B. F. (1957). "Schedules of Reinforcement." Appleton-Century-Crofts, New York.

Fischer, Gloria J. (1962). The formation of learning sets in young gorillas. *J. comp. physiol. Psychol.* **55,** 924.

French, G. M. (1959). Performance of squirrel monkeys on variants of delayed response. *J. comp. physiol. Psychol.* **52,** 741.

French, G. M., & Harlow, H. F. (1962). Variability of delayed-reaction performance in normal and brain-damaged rhesus monkeys. *J. Neurophysiol.* **25,** 585.

Ganz, L., & Riesen, A. H. (1962). Stimulus generalization to hue in the dark-reared macaque. *J. comp. physiol. Psychol.* **55,** 92.

Gardner, L. Pearl, & Nissen, H. W. (1948). Simple discrimination behavior of young chimpanzees: comparisons with human aments and domestic animals. *J. genet. Psychol.* **72,** 145.

Gellermann, L. W. (1931a). The double alternation problem: I. The behavior of monkeys in a double alternation temporal maze. *J. genet. Psychol.* **39,** 50.

Gellermann, L. W. (1931b). The double alternation problem: II. The behavior of children and human adults in a double alternation temporal maze. *J. genet. Psychol.* **39,** 197.

Gellermann, L. W. (1931c). The double alternation problem: III. The behavior of monkeys in a double alternation box-apparatus. *J. genet. Psychol.* **39,** 359.

Ginsburg, N. (1957). Matching in pigeons. *J. comp. physiol. Psychol.* **50,** 261.

Graf, V., & Bitterman, M. E. (1963). General activity as instrumental: Application to avoidance training. *J. exp. Anal. Behav.* **6,** 301.

Graf, V., Bullock, D. H., & Bitterman, M. E. (1964). Further experiments on probability-matching in the pigeon. *J. exp. Anal. Behav.* **7,** 151.

Harlow, H. F. (1949). The formation of learning sets. *Psychol. Rev.* **56,** 51.

Harlow, H. F. (1951). Primate learning. *In* "Comparative Psychology" (C. P. Stone, ed.), 3rd ed., pp. 183–238. Prentice-Hall, New York.

Harlow, H. F. (1958). The evolution of learning. *In* "Behavior and Evolution" (Anne Roe & G. G. Simpson, eds.), pp. 269–290. Yale Univer. Press, New Haven, Connecticut.

Harlow, H. F. (1959). Learning set and error factor theory. *In* "Psychology: A Study of a Science" (S. Koch, ed.), Vol. 2, pp. 492–537. McGraw-Hill, New York.

Harlow, H. F., & Hicks, L. H. (1957). Discrimination learning theory: uniprocess vs. duoprocess. *Psychol. Rev.* **64,** 104.

Harris, J. D. (1943). The auditory acuity of pre-adolescent monkeys. *J. comp. Psychol.* **35,** 255.

Hayes, K. J., Thompson, R., & Hayes, Catherine (1953). Discrimination learning set in chimpanzees. *J. comp. physiol. Psychol.* **46,** 99.

Hilgard, E. R., & Marquis, D. G. (1936). Conditioned eyelid responses in monkeys, with a comparison of dog, monkey, and man. *Psychol. Monogr.* **47,** No. 2 (Whole No. 212), 186–198.

House, Betty J., & Zeaman, D. (1959). Position discrimination and reversals in low-grade retardates. *J. comp. physiol. Psychol.* **52,** 564.

Hull, C. L. (1945). The place of innate individual and species differences in a natural-science theory of behavior. *Psychol. Rev.* **52,** 55.

Hull, C. L. (1952). "A Behavior System." Yale Univer. Press, New Haven, Connecticut.

Hunter, W. S. (1920). The temporal maze and kinesthetic sensory processes in the white rat. *Psychobiology* **2,** 1.

Hunter, W. S. (1928). The behavior of raccoons in a double alternation temporal maze. *J. genet. Psychol.* **35,** 374.

Hunter, W. S., & Nagge, J. W. (1931). The white rat and the double alternation temporal maze. *J. genet. Psychol.* **39,** 303.

Johnson, J. I., Jr. (1961). Double alternation by raccoons. *J. comp. physiol. Psychol.* **54,** 248.

Johnson, J. I., Jr., & Michels, K. M. (1958). Learning sets and object-size effects in visual discrimination learning by raccoons. *J. comp. physiol. Psychol.* **51,** 376.

Jones, L. V. (1954). Distinctiveness of color, form, and position cues for pigeons. *J. comp. physiol. Psychol.* **47,** 253.

Joshi, B. L., & Warren, J. M. (1959). Discrimination of ambivalent cue stimuli by cats. *J. Psychol.* **47,** 3.

Karn, H. W. (1938). The behavior of cats on the double alternation problem in the temporal maze. *J. comp. Psychol.* **26,** 201.

Karn, H. W., & Malamud, H. R. (1939). The behavior of dogs on the double alternation problem in the temporal maze. *J. comp. Psychol.* **27,** 461.

Karn, H. W., & Patton, R. A. (1939). The transfer of double alternation behavior acquired in a temporal maze. *J. comp. Psychol.* **28,** 55.

Kelleher, R. T., & Cook, L. (1959). An analysis of the behavior of rats and monkeys on concurrent fixed-ratio avoidance schedules. *J. exp. Anal. Behav.* **2,** 203.

Kimble, G. A. (1961). "Hilgard and Marquis' Conditioning and Learning," 2nd ed. Appleton-Century-Crofts, New York.

Kirk, K. L., & Bitterman, M. E. (1963). Habit reversal in the turtle. *Quart. J. exp. Psychol.* **15,** 52.

Konorski, J., & Lawicka, W. (1964). Analysis of errors by prefrontal animals on the delayed-response test. *In* "The Frontal Granular Cortex and Behavior" (J. M. Warren & K. Akert, eds.), pp. 271–294. McGraw-Hill, New York.

Koronakos, C., & Arnold, W. J. (1957). The formation of learning sets in rats. *J. comp. physiol. Psychol.* **50,** 11.

Lashley, K. S. (1929). "Brain Mechanisms and Intelligence." Univer. Chicago Press, Chicago, Illinois.

Lashley, K. S. (1938a). Conditional reactions in the rat. *J. Psychol.* **6,** 311.

Lashley, K. S. (1938b). The mechanism of vision: XV. Preliminary studies of the rat's capacity for detail vision. *J. gen. Psychol.* **18,** 123.

Lawicka, W. (1959). Physiological mechanism of delayed reactions: II. Delayed reactions in dogs and cats to directional stimuli. *Acta Biol. exp.* **19,** 199.

Leary, R. W., Harlow, H. F., Settlage, P. H., & Greenwood, D. D. (1952). Performance on double-alternation problems by normal and brain-injured monkeys. *J. comp. physiol. Psychol.* **45,** 576.

Livesey, P. J. (1964). A note on double alternation by rabbits. *J. comp. physiol. Psychol.* **57,** 104.

Maier, N. R. F., & Schneirla, T. C. (1935). "Principles of Animal Psychology." McGraw-Hill, New York.

Mason, W. A., & Harlow, H. F. (1958). Performance of infant rhesus monkeys on a spatial discrimination problem. *J. comp. physiol. Psychol.* **51,** 71.

Meyer, D. R. (1960). The effects of differential probabilities of reinforcement on discrimination learning by monkeys. *J. comp. physiol. Psychol.* **53,** 173.

Meyer, D. R., & Harlow, H. F. (1949). The development of transfer of response to patterning by monkeys. *J. comp. physiol. Psychol.* **42,** 454.

Meyer, D. R., Miles, R. C., & Ratoosh, P. (1954). Absence of color vision in cat. *J. Neurophysiol.* **17,** 289.

Meyers, W. J., McQuiston, M. D., & Miles, R. C. (1962) Delayed-response and learning-set performance of cats. *J. comp. physiol. Psychol.* **55,** 515.

Michels, K. M., & Brown, D. R. (1959). The delayed-response performance of raccoons. *J. comp. physiol. Psychol.* **52,** 737.

Miles, R. C. (1957). Learning-set formation in the squirrel monkey. *J. comp. physiol. Psychol.* **50,** 356.

Miles, R. C. (1958). Color vision in the marmoset. *J. comp. physiol. Psychol.* **51,** 152.

Miles, R. C. (1964). Learning by squirrel monkeys with frontal lesions. *In* "The Frontal Granular Cortex and Behavior" (J. M. Warren & K. Akert, eds.), pp. 149–167. McGraw-Hill, New York.

Miles, R. C. & Meyer, D. R. (1956). Learning sets in marmosets. *J. comp. physiol. Psychol.* **49,** 219.

Nissen, H. W., Riesen, A. H., & Nowlis, V. (1938). Delayed response and discrimination learning by chimpanzees. *J. comp. Psychol.* **26,** 361.

Noble, M., & Adams, C. K. (1963). Conditioning in pigs as a function of the interval between CS and US. *J. comp. physiol. Psychol.* **56,** 215.

Noble, M., & Harding, G. E. (1963). Conditioning in rhesus monkeys as a function of the interval between CS and US. *J. comp. physiol. Psychol.* **56,** 220.

Noble, M., Gruender, Anne, & Meyer, D. R. (1959). Conditioning in fish (*Mollienisia* sp.) as a function of the interval between CS and US. *J. comp. physiol. Psychol.* **52,** 236.

Noer, Mary C., & Harlow, H. F. (1946). Discrimination of ambivalent cue stimuli by macaque monkeys. *J. gen. Psychol.* **34,** 165.

North, A. J., Maller, O., & Hughes C. (1958). Conditional discrimination and stimulus patterning. *J. comp. physiol. Psychol.* **51,** 711.

Orbach, J. (1959). Disturbances of the maze habit following occipital cortex removals in blind monkeys. *A. M. A. Arch. Neurol. Psychiat.* **81,** 49.

Pastore, N. (1954). Discrimination learning in the canary. *J. comp. physiol. Psychol.* **47,** 389.

Rabinovitch, M. S., & Rosvold, H. E. (1951). A closed-field intelligence test for rats. *Canad. J. Psychol.* **5**, 122.

Razran, G. (1961). The observable unconscious and the inferrable conscious in current Soviet psychophysiology: Interoceptive conditioning, sematic conditioning, and the orienting reflex. *Psychol. Rev.* **68**, 81.

Riopelle, A. J., & Copelan, E. L. (1954). Discrimination reversal to a sign. *J. exp. Psychol.* **48**, 149.

Rollin, A. R. (1963). Successive object discrimination reversals by squirrels. Paper read at East. Psychol. Ass., New York.

Rosenzweig, M. R., Krech, D., & Bennett, E. L. (1960). A search for relations between brain chemistry and behavior. *Psychol. Bull.* **57**, 476.

Rumbaugh, D. M., & Rice, Carol P. (1962). Learning-set formation in young great apes. *J. comp. physiol. Psychol.* **55**, 866.

Schlosberg, H., & Katz, A. (1943). Double alternation lever-pressing in the white rat. *Amer. J. Psychol.* **56**, 274.

Schusterman, R. J. (1962). Transfer effects of successive discrimination-reversal training in chimpanzees. *Science* **137**, 422.

Schusterman, R. J., & Bernstein, I. S. (1962). Response tendencies of gibbons in single and double alternation tasks. *Psychol. Rep.* **11**, 521.

Shell, W. F., & Riopelle, A. J. (1957). Multiple discrimination learning in raccoons. *J. comp. physiol. Psychol.* **50**, 585.

Shell, W. F., & Riopelle, A. J. (1958). Progressive discrimination learning in platyrrhine monkeys. *J. comp. physiol. Psychol.* **51**, 467.

Sidman, M. (1953a). Avoidance conditioning with brief shock and no exteroceptive warning signal. *Science* **118**, 157.

Sidman, M. (1953b). Two temporal parameters of the maintenance of avoidance behavior by the white rat. *J. comp. physiol. Psychol.* **46**, 253.

Sidman, M. (1955). Some properties of the warning stimulus in avoidance behavior. *J. comp. physiol. Psychol.* **48**, 444.

Sidman, M., & Boren, J. J. (1957). A comparison of two types of warning stimulus in an avoidance situation. *J. comp. physiol. Psychol.* **50**, 282.

Skinner, B. F., (1950). Are theories of learning necessary? *Psychol. Rev.* **57**, 193. (Reprinted in "Cumulative Record," pp. 39–69. Appleton-Century-Crofts, New York, 1959).

Skinner, B. F. (1956). A case history in scientific method. *Amer. Psychologist* **11**, 221. (Reprinted in "Cumulative Record," pp. 76–100. Appleton-Century-Crofts, New York, 1959. Also in "Psychology: A Study of a Science" [S. Koch, ed.], Vol. 2, pp. 359–379. McGraw-Hill, New York, 1959.)

Spence, K. W. (1942). The basis of solution by chimpanzees of the intermediate size problem. *J. exp. Psychol.* **31**, 257. (Reprinted in "Behavior Theory and Learning," pp. 339–358. Prentice-Hall, Englewood Cliffs, New Jersey, 1960.)

Stewart, C. N., & Warren, J. M. (1957). The behavior of cats on the double-alternation problem. *J. comp. physiol. Psychol.* **50**, 26.

Sutherland, N. S. (1961). The methods and findings of experiments on the visual discrimination of shape by animals. *Exp. Psychol. Soc. Monogr.* **1**, 1.

Tyrrell, D. J. (1963). The formation of object discrimination learning sets by rats. Paper read at East. Psychol. Ass., New York.

Voronin, L. G. (1962). Some results of comparative-physiological investigations of higher nervous activity. *Psychol. Bull.* **59**, 161.

Warren, J. M. (1959a). Solution of object and positional discriminations by rhesus monkeys. *J. comp. physiol. Psychol.* **52**, 92.

Warren, J. M. (1959b). Stimulus perserveration in discrimination learning by cats. *J. comp. physiol. Psychol.* **52**, 99.

Warren, J. M. (1960a). Discrimination reversal learning by cats. *J. genet. Psychol.* **97**, 317.

Warren, J. M. (1960b). Oddity learning set in a cat. *J. comp. physiol. Psychol.* **53**, 433.

Warren, J. M. (1960c). Reversal learning by paradise fish (*Macropodus opercularis*). *J. comp. physiol. Psychol.* **53**, 376.

Warren, J. M. (1960d). Solution of sign-differentiated object and positional discriminations by rhesus monkeys. *J. genet. Psychol.* **96**, 365.

Warren, J. M. (1960e). Supplementary report: Effectiveness of food and nonfood signs in reversal learning by monkeys. *J. exp. Psychol.* **60**, 263.

Warren, J. M. (1961). Individual differences in discrimination learning by cats. *J. genet. Psychol.* **98**, 89.

Warren, J. M. (1964a). The behavior of carnivores and primates with lesions in the prefrontal cortex. *In* "The Frontal Granular Cortex and Behavior" (J. M. Warren & K. Akert, eds.), pp. 168–191. McGraw-Hill, New York.

Warren, J. M. (1964b). Additivity of cues in conditional discrimination learning by rhesus monkeys. *J. comp. physiol. Psychol.* **58**, 124.

Warren, J. M., & Baron, A. (1956). The formation of learning sets by cats. *J. comp. physiol. Psychol.* **49**, 227.

Warren, J. M., & Sinha, M. M. (1959). Interactions between learning sets in monkeys. *J. genet. Psychol.* **95**, 19.

Warren, J. M., & Warren, Helen B. (1962). Reversal learning by horse and raccoon. *J. genet. Psychol.* **100**, 215.

Warren, J. M., Leary, R. W., Harlow, H. F., & French, G. M. (1957). Function of association cortex in monkeys. *Brit. J. Anim. Behav.* **5**, 131.

Warren, J. M., Brookshire, K. H., Ball, G. G., & Reynolds, D. V. (1960). Reversal learning by white leghorn chicks. *J. comp. physiol. Psychol.* **53**, 371.

Warren, J. M., Warren, Helen B., & Akert, K. (1961). *Umweg* learning by cats with lesions in the prestriate cortex. *J. comp. physiol. Psychol.* **54**, 629.

Warren, J. M., Warren, Helen B., & Akert, K. (1962). Orbitofrontal cortical lesions and learning in cats. *J. comp. Neurol.* **118**, 17.

Warren, J. M., Grant, R., Hara, K., & Leary, R. W. (1963). Impaired learning by monkeys with unilateral lesions in association cortex. *J. comp. physiol. Psychol.* **56**, 241.

Weaver, L. A., Jr., & Michels, K. M. (1961). Methodological factors affecting the formation of learning sets by rats. *Anim. Behav.* **9**, 4.

Wilson, W. A., Jr. (1960). Supplementary report: Two-choice behavior of monkeys. *J. exp. Psychol.* **59**, 207.

Wilson, W. A., Jr., & Rollin, A. R. (1959). Two-choice behavior of rhesus monkeys in a noncontingent situation. *J. exp. Psychol.* **58, 174.**

Wilson, W. A., Jr., Oscar, Marlene, & Bitterman, M. E. (1964). Probability-learning in the monkey. *Quart. J. exp. Psychol.* **16**, 163.

Wodinsky, J., & Bitterman, M. E. (1953). The solution of oddity-problems by the rat. *Amer. J. Psychol.* **66**, 137.

Wright, P. L., Kay, H., & Sime, M. E. (1963). The establishment of learning sets in rats. *J. comp. physiol. Psychol.* **56**, 200.

Yamaguchi, S., & Warren, J. M. (1961). Single versus double alternation learning by cats. *J. comp. physiol. Psychol.* **54**, 533.

Zeigler, H. P. (1961). Learning-set formation in pigeons. *J. comp. physiol. Psychol.* **54,** 252.

Zimmerman, J., & Ferster, C. B. (1963). Intermittent punishment of SΔ responding in matching to sample. *J. exp. Anal. Behav.* **6,** 349.

Zimmermann, R. R. (1963). The performance of baby monkeys on a closed-field intelligence test for rats. Paper read at East. Psychol. Ass., New York.

Appendix

As the chapters of this book amply demonstrate, not all primates behave alike. Different species, and even different subspecies, may differ in behavior. Although this is widely recognized, an important implication for behavioral research is often ignored. Because the results of a study may depend on the particular primate being studied, it is imperative that the researcher identify his subjects as fully and accurately as possible.

All too often, only a vernacular name is mentioned in the report of a behavioral investigation. But a vernacular name is not adequate identification; usage varies from time to time. This variation is reflected in a number of differences between vernacular names used in this book and those used when the studies being described were originally published.

In this book, the first time that a particular vernacular name appears in each chapter, the scientific name of the animal is also given. Often only the genus can be named; this further reflects inadequate identification in the original publication.

Even when a Latin name was given in the original study, the same subjects may be identified with a different Latin name in this book. Such differences sometimes arise from changes in accepted nomenclature, but perhaps more often they result from the marked differences among primate taxonomists in their systems of classification and nomenclature. Any exhortation to use scientific names in identifying primates must be tempered by the knowledge that there is no single authoritative source for primate nomenclature. Fortunately, cases in which the same name is used for different species are now quite rare, but the literature often contains different names for the same species.

To avoid such inconsistencies within the covers of this book, the scientific names used are those given by Fiedler (1956). Fiedler's nomenclature was adopted because it is recent and comprehensive, and also because it conforms in general to the scientific nomenclature familiar to behavioral scientists. An example of a recent and comprehensive system that differs markedly from familiar usage is that of Sanderson (1957), in which, for example, the macaques (usually considered a single genus, *Macaca*) are classified as a subfamily and split into eight genera.

For the benefit of readers who would like to see how various genera are related to the rest of the Order Primates, or who want an overview

TABLE I

SOME PROBLEMS OF PRIMATE NOMENCLATURE

Scientific name (Fiedler, 1956)	Vernacular name(s)	Other Latin names sometimes used
Callithrix	Marmoset	*Hapale*
Saimiri sciureus	Squirrel monkey	*Saimiri sciurea*
Cebus albifrons	White-fronted capuchin Cinnamon ringtail	
Cebus apella	Hooded capuchin Large-headed capuchin Tufted capuchin	*Cebus fatuellus* (for hooded capuchin)
Macaca irus[a]	Cynomolgus macaque Crab-eating macaque Java monkey	*Macaca fascicularis* *Macaca mulatta fascicularis*
Papio comatus	Chacma baboon	*Papio ursinus* *Papio porcarius*
Papio cynocephalus	Yellow baboon Sphinx baboon	*Papio sphinx*[b]
Papio doguera	Anubis baboon Doguera baboon Olive baboon[c] Guinea baboon[c]	*Papio anubis*
Papio leucophaeus	Drill	*Mandrillus leucophaeus*
Papio sphinx	Mandrill	*Mandrillus sphinx*
Cercocebus torquatus atys	Sooty mangabey	*Cercocebus fuliginosus*
Cercopithecus aethiops	Green monkey Vervet Grivet	*Cercopithecus sabaeus* *Cercopithecus pygerythrus*
Presbytis entellus	Indian langur Hanuman langur Sacred langur	*Semnopithecus entellus*
Presbytis johni	Nilgiri langur	*Kasi johni*
Presbytis senex	Purple-faced langur	*Kasi vetulus*
Pan troglodytes	Chimpanzee	*Pan satyrus*

[a]Fooden (1964) claims that "*irus*" is not a valid name and should be replaced by "*fascicularis*." He also claims that the cynomolgus macaque and rhesus macaque are conspecific. If both of these claims are true, the cynomolgus macaque is *Macaca mulatta fascicularis* and the rhesus macaque is *Macaca mulatta mulatta*. There have been other recent indications that all macaques belong to a single species, and there have also been hints that baboons and macaques should be classified in the same genus.

[b]This is the name used by Fiedler (1956) for the mandrill.

[c]*Papio papio*, not listed in this table, is also called "olive baboon" and "Guinea baboon" as well as "western baboon."

of the whole order, a classification of the living primates is presented on the endpapers of each volume. The classification is essentially that of Fiedler (1956); the sequence is changed slightly to list the Callithricidae before the Cebidae (and the Callimiconinae before the Cebinae), and Fiedler's Infraorders Tupaiiformes, Lemuriformes, Lorisiformes, and Tarsiiformes are listed as the Superfamilies Tupaioidea, Lemuroidea, Lorisoidea, and Tarsioidea. Probably none of these admittedly arbitrary changes would seriously bother even a taxonomist; most of the changes, in fact, are in accord with classifications by Simpson (1945) and by Straus (1949), which are in other respects similar to Fiedler's.

The use of a consistent nomenclature is necessary, but it doesn't eliminate alternative names from the literature. Some readers may be familiar with names from other nomenclatures, and they may fail to recognize a familiar species discussed under an unfamiliar name. This is one reason for using vernacular names throughout this book in addition to scientific names. A few possible sources of confusion are listed in Table I.

References

Fiedler, W. (1956). Übersicht über das System der Primates. *In* "Primatologia" (H. Hofer, A. H. Schultz, & D. Starck, eds.), Vol. I, pp. 1–266. Karger, New York.

Fooden, J. (1964). Rhesus and crab-eating macaques: intergradation in Thailand. *Science* **143**, 363.

Sanderson, I. T. (1957). "The Monkey Kingdom." Hanover House, Garden City, New York.

Simpson, G. G. (1945). The principles of classification and a classification of mammals. *Bull. Amer. Mus. nat. Hist.* **85**, 1.

Straus, W. L., Jr. (1949). The riddle of man's ancestry. *Quart. Rev. Biol.* **24**, 200.

Author Index

Numbers in italic indicate the pages on which the complete references are listed.

D

E

F

G

H

Subject Index

(Volume I: pages 1–285; Volume II: pages 287–595)

R

Summary of Primate Classification

Suborder	Superfamily	Family	Subfamily	Genus	Vernacular name(s)
Prosimiae (prosimians)	Tupaioidea	Tupaiidae	Tupaiinae	*Tupaia* *Dendrogale* *Urogale*	Tree shrew Smooth-tailed tree shrew Philippine tree shrew
			Ptilocercinae	*Ptilocercus*	Pen-tailed tree shrew
	Lemuroidea	Lemuridae	Lemurinae (greater lemurs)	*Lemur* *Hapalemur* *Lepilemur*	Lemur Gentle lemur Sportive lemur
			Cheirogaleinae (lesser lemurs)	*Cheirogaleus* *Microcebus*	Dwarf lemur, mouse lemur
		Indriidae	Indriinae	*Avahi* *Propithecus* *Indri*	Woolly lemur Sifaka Indri
		Daubentoniidae		*Daubentonia*	Aye-aye
	Lorisoidea	Lorisidae		*Loris* *Nycticebus* *Arctocebus* *Perodicticus*	Slender loris Slow loris Angwantibo Potto
		Galagidae		*Galago*	Galago (bush-baby)
	Tarsioidea	Tarsiidae		*Tarsius*	Tarsier
		Callithricidae		*Callithrix* *Leontocebus*	Marmoset Tamarin, pinche
			Callimiconinae	*Callimico*	Goeldi's "marmoset"
				Aotes	Douroucouli (night monkey,